DEGRADATION, MITIGATION, AND FORECASTING APPROACHES IN THIN FILM PHOTOVOLTAICS

DEGRADATION, MITIGATION, AND FORECASTING APPROACHES IN THIN FILM PHOTOVOLTAICS

DIPANKAR DEB
Department of Electrical Engineering and Computer Science
Institute of Infrastructure, Technology, Research and Management
Ahmedabad, India

KSHITIJ BHARGAVA
Department of Electrical Engineering and Computer Science
Institute of Infrastructure, Technology, Research and Management
Ahmedabad, India

ELSEVIER

ACADEMIC PRESS
An imprint of Elsevier

Academic Press is an imprint of Elsevier
125 London Wall, London EC2Y 5AS, United Kingdom
525 B Street, Suite 1650, San Diego, CA 92101, United States
50 Hampshire Street, 5th Floor, Cambridge, MA 02139, United States
The Boulevard, Langford Lane, Kidlington, Oxford OX5 1GB, United Kingdom

Library of Congress Cataloging-in-Publication Data
A catalog record for this book is available from the Library of Congress

British Library Cataloguing-in-Publication Data
A catalogue record for this book is available from the British Library

ISBN: 978-0-12-823483-9

For information on all Academic Press publications
visit our website at https://www.elsevier.com/books-and-journals

Publisher: Joe Hayton
Acquisitions Editor: Lisa Reading
Editorial Project Manager: Michelle Fisher
Production Project Manager: Prasanna Kalyanaraman
Designer: Matthew Limbert

Typeset by VTeX

Dedicated to Vivasvān for being the secondary source of all the energy that we get to use.

-Dipankar Deb

Dedicated to my father Sanjay Kumar Bhargava, my mother Navita Bhargava, my wife Kshamata Bhargava, and my son Rakshit Bhargava.

-Kshitij Bhargava

Contents

List of figures

List of tables

Biography

Dipankar Deb is a Senior Member of IEEE and served a couple of years at IIT Guwahati as an assistant professor (AGP 8000) during 2010–2012. He has over six years of industrial experience in both New York (USA) and GE Global Research (Bengaluru, India). From July 2015 to January 2019, he served as an associate professor, and since 24 January 2019 he is Professor in Electrical Engineering at the Institute of Infrastructure Technology Research and Management (IITRAM) Ahmedabad. He holds six US patents and one India Patent, and he has published 36 SCI-indexed journal articles and more than 40 international conference papers. He has also authored ten books with reputed publishers like Elsevier and Springer. He is book series editor with CRC Press for Control Theory and Applications and is associate editor of IEEE Access. He is also a book series editor of "Studies in Infrastructure and Control" at Springer. He has also worked extensively in areas such as adaptive control, active flow control, renewable energy, cognitive robotics, and machine learning.

Kshitij Bhargava received a B.Tech. degree in Electrical and Electronics Engineering from Uttar Pradesh Technical University, Lucknow, India (2009), an M.Tech. degree in Microelectronics Systems and Embedded Technology from Jaypee Institute of Information Technology, Noida, India (2011), and a Ph.D. degree in Electrical Engineering from Indian Institute of Technology Indore, India (2016). Since 2016, Dr. Bhargava has been working as assistant professor in the Department of Electrical Engineering, Institute of Infrastructure, Technology, Research And Management, Ahmedabad, India. Dr. Bhargava has published in total of 14 SCI-indexed publications in reputed scientific journals, two book chapters, and more than 20 conference publications to date. His research interests include organic electronics, field-effect transistors, thin-film characterizations, flexible electronics, solar cells, and VLSI design.

Preface

High insolation in arid regions of the world reduces solar energy costs and provides a pivotal motivation to deploy solar panels. There are also several challenges, and one needs to know what effect temperature and dust have on the solar cells and PV modules. This book looks at the cleaning methods used to reduce soiling, the measurement of solar cell and PV module performance, and optimization of the technology's design to enhance the energy yield. Consequently, one has the opportunity to consider how to deploy PV technology in a specific context. The heat and dust pose a challenge for PV modules' optimal performance and the applications' reliability. This book deals with mitigating such issues by understanding solar-radiation components and solar cells' characteristics. The reader also learns about forecasting and the influences on selection of a site. The economic impact of PV applications is a crucial area of interest, and this book provides knowledge on the costs and benefits of solar energy from PV applications.

Especially in hot desert climates, one needs to understand better the challenge to address critical knowledge points relevant to the successful delivery of PV applications in harsh environments. Such a process would involve analyzing site-specific surface insolation data to accurately estimate a proposed facility's annual energy production. Everything that impacts how much solar radiation arrives at the surface is essential to consider. In arid climates, this would also include atmospheric aerosols, namely dust and particulates, rain, humidity, and cloud cover, if any, and any number of other site-specific considerations. We also discuss how to identify the best site for our PV application, how to forecast energy production, and how to calculate the cost of PV in desert environments, and we'll explore techno-economic models for PV technology diffusion in these environments. This book will guide the reader by identifying the solar spectrum components and their relevance to solar energy deployment in such environments. We also address atmospheric aerosols' interaction with solar radiation and introduce the reader to short-term forecasting of surface insolation. We also cover the artificial intelligence and neural network applications used for prediction. We also lead the reader by calculating the energy cost from PV and explore techno-economic models of PV technology diffusion in fossil-fuel-rich countries.

The book primarily deals with degradation and mitigation techniques, including forecasting approaches, especially in thin-film photovoltaics. The chapters of this book are organized as follows: Chapter 1 presents discussion pertaining to global energy requirements and the emergence of photovoltaics as renewable energy technology. Chapter 2 describes the process of photovoltaic energy conversion and various conventional and emerging absorber materials for thin-film photovoltaics. Chapter 3 is dedicated to understanding of issues causing performance degradation in thin-film photovoltaics in cells and modules.

Chapter 4 provides details about various methods for mitigating soiling of solar panels, followed by Chapter 5 which provides a water-free fully automated solution of energy-efficient dust removal. Chapter 6 explains the phenomena of potential induced degradation (PID) in thin- film photovoltaics and simulation methodology of the same in various types of thin-film photovoltaic cells. Chapter 7 extends the study of PID through multiple forecasting methods to preempt the loss of the potential yield due to PID in multicrystalline panels, followed by a study of PID-detection techniques in Chapter 8 which, when applied along with forecasting methods, can provide a more accurate level of yield reduction. Chapter 9 is the concluding chapter of this book offering insights into features of the next-generation photovoltaics and issues hindering their progress towards commercialization.

It is hoped that this book will provide the reader with a comprehensive resource for degradation, mitigation and forecasting related issues when it comes to PVs, especially in the arid regions of the world.

Dipankar Deb
Kshitij Bhargava

Acknowledgements

The completion of this book in its present form would not have been feasible without the assistance of the Institute of Infrastructure, Technology, Research and Management, Ahmedabad in providing the necessary infrastructural support. The authors are also grateful to Elsevier for allowing us to use related content for the book. The authors are grateful to the team of engineers from Larsen & Toubro (L&T) Ltd.'s plant situated in Nimbahera, Rajasthan, India for the provision of data and all possible support to our student Mr. Shubham Mishra who painstakingly collected all the relevant data. Author Kshitij Bhargava is grateful to Dr Marc Burgelman, Department of Electronics and Information Systems (ELIS), University of Gent, Belgium, for providing the SCAPS-1D simulation software.

Acronyms

ANN	Artificial Neural Networks
AR	Auto-Regression
ARC	Anti-Reflective Coating
ARIMA	Auto-Regressive Integrated Moving Average
a-Si	Amorphous Silicon
BSF	Back Surface Field
CdTe	Cadmium Telluride
CIGS	Copper Indium Gallium Diselenide
CZTS	Copper Zinc Tin Sulphide
EDS	Electro-Dynamic Screen
ELECTRE	Elimination Et Choix Traduisant la Realité
EVA	Ethylene-Vinyl-Acetate
FF	Fill Factor
GaAs	Gallium Arsenide
GBM	Gradient Boosted Machines
GDP	Gross Domestic Product
GHG	Greenhouse Gas
IRENA	International Renewable Energy Agency
I-V	Current-Voltage
LCOE	Levelized Cost of Energy
PID	Potential Induced Degradation
PLA	Persistent Learning Algorithm
PV	Photovoltaic
MA	Moving Average
MCDM	Multi-Criteria Decision Making
MSE	Mean Squared Error
QE-λ	Quantum Efficiency Wavelength
RF	Random Forest
RMSE	Root-Mean-Square Error
RSE	Residual Squared Error
SAW	Simple Additive Weighting
SCAPS	Solar-Cell Capacitance Simulator
SES	Single Exponential Smoothing
SHIP	Super-Hydrophilic Plane
SHOP	Super-hydrophobic Plane
SLG	Soda Lime Glass
SRH	Shockley–Read–Hall
SVM	Support Vector Machines
SVR	Support Vector Regression
STC	Standard Test Conditions
TFPV	Thin-Film Photovoltaics
TOPSIS	Technique for Order of Preference by Similarity to Ideal Solution
UV	Ultra Violet

CHAPTER 1

Introduction to photovoltaics

The requirement for renewable energy sources is growing day by day. In this context, solar energy and wind energy are the most promising technologies owing to their consistent availability. In particular, solar-energy harnessing has grabbed great attention in recent times. Although there are two basic techniques to harness solar energy, namely, solar-thermal and solar photovoltaic, the latter has proven to be the more efficient and useful one. For large-scale generation of electricity, solar photovoltaic panels are combined into an array. This chapter presents a detailed overview of solar energy-harnessing technology's status, emphasizing global warming, and photovoltaic manufacturing. Additionally, the fundamentals of photovoltaic operation are also discussed.

1.1 Introduction

The energy requirement of a nation is crucial for its socioeconomic development. Most energy needs are currently fulfilled through fossil fuel, which has led to severe environmental consequences like global warming and climate change. Therefore, as per the UN Framework, a dedicated international commitment is needed to address the aforesaid issue by ensuring the fulfillment of continuously escalating energy requirements. Since the Oil Crisis of 1973 [1], research to harness energy from renewable sources has grown greatly in the subsequent decades. Among several available renewable sources, solar energy is one such resource that is most abundant, clean, and inexhaustible [2]. Like any other renewable energy source, solar energy is available free of cost for a sufficient duration in a day. However, harnessing the energy of the sun requires an advanced scientific technology popularly known as photovoltaic technology. The research and development of photovoltaic cells and modules have gone through several commercialization phases in recent decades [3]. This chapter presents an overview of solar photovoltaic technology. It will enable the readers to understand the requirement of photovoltaic technology and the intricacies of its development for successful commercialization. In addition, we will cover the relevant basic concepts briefly to explain the subsequent chapters better.

Degradation, Mitigation, and Forecasting Approaches in Thin Film Photovoltaics
https://doi.org/10.1016/B978-0-12-823483-9.00012-7

1.2 Scenario of global energy requirements

The current scenario for meeting the continuously increasing energy needs is mainly fulfilled through fossil fuels and some other conventional resources (e.g., nuclear and hydro) [4]. According to the recently available data, the global energy requirements rose by 2.3% in 2018, which is the fastest in the current decade. Although the solar and wind-energy generation doubled in this era, most of the requirements are still getting fulfilled through fossil fuels (70%). For significant gross domestic product (GDP) growth every year, both developed and developing countries' energy requirements are bound to increase. Also, the world's growing population is causing a sharp rise in global energy needs. Therefore, it becomes necessary to present an accurate scenario for the worldwide energy requirements in the coming decades.

The prediction of the global energy scenario is strongly dependent on geography, population, technology, government policies, climate change, etc. These must be taken into account for accurate predictions. The motive behind such predictions is to channel the planning required to reduce carbon dioxide (CO_2) emissions and achieve environmental sustainability [5]. In this context, electrification through renewable sources offers a clean and green alternative to the problem by creating prospects of reducing air pollution, improving health standards, and supporting overall socioeconomic development. In pursuit of achieving a sustainable environment, the world's major economies are taking significant, although a still insufficient number of initiatives (like subsidies) to commercialize electricity as the primary secondary source of energy by 2050. With technological advancements, it would be even possible to supply more than 80% of the global power demand through renewable sources. Therefore, a significant part of the GDP needs to be invested in technological growth. A famous example of such technology is the emergence of electric vehicles, which will be completely powered by the electric charge stored in highly efficient batteries.

The present status of CO_2 content in the atmosphere has reached alarming levels, and it would be suicidal not to address the situation with dedicated efforts. We expect that a nearly 70% reduction in the global CO_2 emissions would be desirable to avoid the catastrophic consequences of global warming, achievable with an annual emission reduction of 3.5%. Therefore, we must address the critical technological advancements, combined with awareness about the environmental conservation, thoughtfully. A well-planned transitional road map based on the trade-off between socioeconomic growth and climate conservation would serve the purpose.

However, the geographic conditions and population explosion in major economies will be severe bottlenecks. The elusive benefit of forecasting the global energy requirement will lead to the conservation of precious fossil-fuel reserves and also open the window to significant employment growth in the sector of renewable energy electrification.

The technology needed to harness the energy from naturally available nonconventional renewable sources has seen many decades of extensive research and development. Current efforts aim towards developing technology for renewable electrification to the level of large-scale and cost-effective deployment. One of the major hindrances to this process is the unreliable and deficient governmental policies across different nations that need to evolve through much more coordinated policy-making and setting target-driven efforts. The seriousness of the matter is evident from the fact that, despite increased awareness, support for initiatives like the Paris Agreement on climate change, and the emergence of clean and sustainable energy alternatives, CO_2 emissions worldwide have increased 1.3% annually over the past five years. Such an increase signifies a wide gap between our initiatives and the outcomes. The potential of renewable energy-based electrification infrastructure, if established successfully, can achieve up to 75% reductions in global CO_2 emissions.

Such an objective will require a two-third's share of renewable electrification by 2050 to the currently existing one-sixth. As per the current and futuristic policies, the estimated investment of USD 95 trillion would be require up to 2050, which is 2% of global GDP. This would be a severe burden on countries in the low-income group and thus requires adequate planning. A forecast scenario for rising renewable-energy share is depicted in Fig. 1.1.

1.3 Global warming and renewable energy

Global warming refers to the rise in the average temperature of the earth's surface due to the greenhouse effect. Since the start of the Industrial Revolution, the overuse of fossil fuels reserves has led to the increasing content of greenhouse gases (GHGs), like CO_2, water vapor, methane, ozone, chlorofluorocarbons (CFCs), and nitrous oxide [6], in the atmosphere. GHGs absorb and thereby trap the infrared radiation that maintains the Earth's temperature. Among GHGs, the concentration of CO_2 acts as the major controlling parameter of Earth's temperature and makes it habitable for life. However, beyond the permissible limits, the levels of CO_2 can significantly

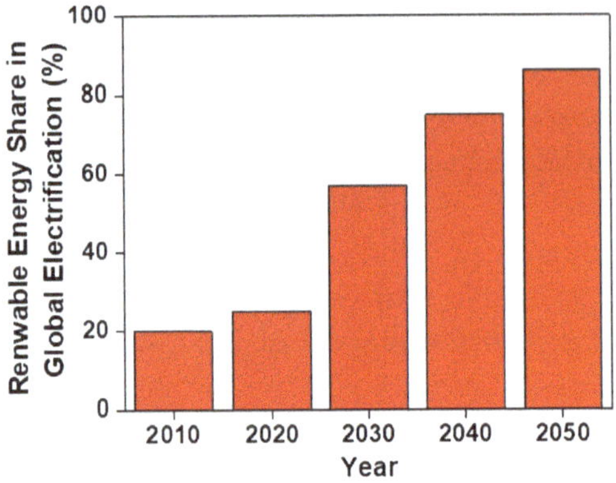

Figure 1.1 Road map of existing and rising percentage share of renewable energy in global electrification.

increase the temperature of Earth, resulting in dire consequences. Research studies predict that increasing the CO_2 levels to double what is required can increase the temperature of Earth by 3.8°C. Therefore, a well-planned strategy is needed to counter the growing problem of global warming, while ensuring sufficient socioeconomic growth at the same time. The content of CO_2 in Earth's atmosphere has reached alarming levels, and the days are not far away when we can expect some catastrophic consequences like extinction of glaciers feeding perennial rivers and rising sea levels submerging the island nations like the Maldives if not addressed soon. Eye-opening evidence is the fact that 2018 was the fourth warmest year among the 20 warmest years that have also happened in last 22 years, with the top four being the last four years. The average global temperature has increased 1.04°C above the normal during 2014-2018.

The CO_2 concentration in the atmosphere has sharply increased by 30% since 1950, and main reason for this has been the burning of fossil fuels (especially coal) to produce heat and electricity, combined with deforestation as trees are responsible for the absorption of carbon. Post the Industrial Revolution, the temperature of the Earth has undergone drastic rise due to the burning of fossil fuels and deforestation, and according to latest scientific prediction, it will rise 2°C to 6°C further by the end of 21st century. The concentration of CO_2 takes a few to hundreds of years to balance itself in the atmosphere. Apart from CO_2, water vapor is also an important component in climate change because it constitutes two-thirds of the global warming and thus plays a critical role with CO_2 in balancing the temperature of the Earth. The extent of warming of Earth's climate is studied through the content of water vapor present in the atmosphere, which seeks to maintain an equilibrium between temperature and water vapor. With the rise in temperature due to increasing CO_2 emissions, this balance gets disturbed, and climate warming happens at a faster pace because more water-vapor content adds to the heat entrapment in the atmosphere.

After water vapor and CO_2, the second largest GHG component is methane in the environment. The content of methane has doubled in the past 150 years, and 60% of it has been due to the human activities. The primary source of methane is the extraction, transportation, refining, and distribution of natural gas, while agricultural waste, human waste, and landfills are the second largest source of methane. Although methane has a greater capability to absorb and reradiate heat, CO_2 has been considered more responsible for global warming because its concentration and lifetime is 200-times more than methane. Apart from these, N_2O also contributes to global temperature rise, and it comes from fossil-fuel combustion, agricultural and industrial processes, and waste-water treatment. An important characteristic of N_2O is that it is retained in the atmosphere for more than 100 years and can only be removed through the nitrogen cycle by the action of bacteria, certain chemical reactions, and UV radiations. Although N_2O has a 300-times greater capability to entrap heat in the atmosphere, its is hardly considered critical for global warming because its concentration is much lower than CO_2. Fig. 1.2 shows the current scenario of GHG emission in which the levels of CO_2 has increased beyond safe levels, causing global warming. The consequences of global warming could be disastrous and widespread.

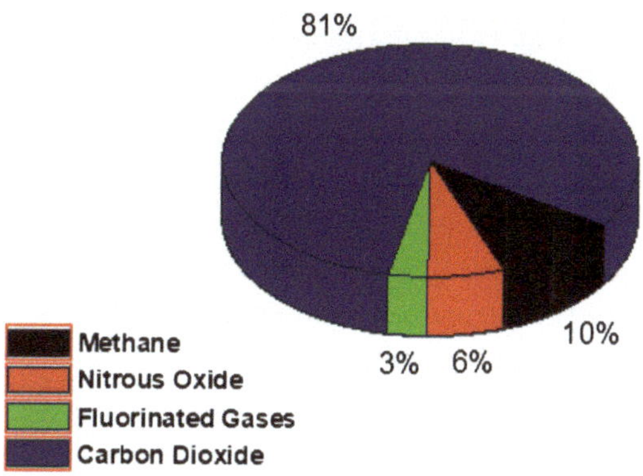

Figure 1.2 Overview of greenhouse-gas emission in 2017.

The continuously increasing levels of GHGs, and especially CO_2, have led to the evident effects of global warming, such as the shift in climatic conditions, crop yields, frequent droughts and floods, the melting of glaciers, rising acidity of sea water affecting marine life, among others. Also, the majority of the heat trapped by GHGs is absorbed by ocean water, leading to its rise in temperature. This increased temperature reduces the content of dissolved oxygen and makes the habitat less favorable for marine life (e.g., diminishing algae and coral lives) and also causing water expansion that leads to a rising level (a 2-m rise in sea level will submerged many coastal cities). Moreover, a rising difference in temperature between the poles and Equator will enhance the frequency of cyclones and hurricanes.

Apart from frequent droughts, floods and heat waves are some of the other noticeable consequences of global warming. Therefore, renewable-energy systems such solar photovoltaics (PVs), wind-energy mills, bio-fuels, and hydropower automatically become the need of the hour. The importance of renewable energy grows many times considering the various aspects of civilization, such as socioeconomics, energy access, energy security, reduction in climate change, and environmental and health impacts. The climate of Earth has already started to show the signs of saturation, and therefore it is necessary to significantly reduce our dependence on fossil fuels and initiate the commercialization of clean and green renewable-energy technology on a massive scale.

One of the primary motivating factors of photovoltaic adoption is that we are already adopting photovoltaic technologies quite rapidly. And we see an expansion of utility-scale solar happening, as well. So, why are we doing that? One of the reasons, you might think, would be economics. But there are many other competing technologies to solar energy to produce electricity—in this case, photovoltaics‡that are lower cost. And if we were to use purely market-based decision criteria, we choose those other technologies. So, something else is motivating us. To a large extent, that motivation is our concern for Earth's environment. We know that we have a shared relationship with our environment. We make decisions to adopt certain technologies because we see them as technologies that will help ensure a longer, more sustainable future in terms of our energy use.

We've already devised some fascinating strategies to reduce our energy use. We've worked hard to adopt new technologies, new lighting strategies, and new energy-efficiency strategies that reduce our electricity demand. Photovoltaics has an exciting structure that starts with a module, and the modules are grown in pieces, so it's very flexible for different sizes and scales. So, it allows us to adopt a technology that has a route to reducing our carbon intensity; it can enable us, as a society, to have greater energy independence. There are a lot of exciting incentives that are all looping around. We have positive feedback because, as we choose to adopt solar energy, we ultimately see a financial return on investment. But we also see a return on energy. Moreover, we see a return in lower emissions that allows us to reduce the risk of societal growth having a significant impact on climate change through greenhouse-gas emissions. Because it will enable us to fulfill both kinds of obligations, societal and environmental, across the globe and for future generations in a way that we can feel very comfortable with, we look forward to a future in solar energy.

Renewable energy systems have emerged as an important pillar of climate protection and sustainable economic development. Supported by the technological advancements and favorable governmental policy changes, the commercial costs of renewable energy systems are decreasing continuously and also creating a plethora of employment prospects in this sector. Foreseeing the possible devastations of global warming, the world is now moving towards low-carbon economic growth. According to the data assessment by the International Renewable Energy Agency (IRENA), around 11 million jobs are expected to be produced in the renewable energy sector, including bioenergy [7] and wind energy [8], with 3.9 million jobs alone in solar PV [9,10] manufacturing. According to the IRENA data shown in Fig. 1.3, it is important to notice that since 2010 there has been sharp rise in the employment prospects in the renewable energy sector with solar PV being in the top position.

Bioenergy includes liquid biofuels, solid biomass, and biogas, while other technologies consists of geothermal energy, concentrated solar power, heat pumps, municipal and industrial waste, and ocean energy.

1.4 Brief overview of PV manufacturing

What have been the most serious bottlenecks in the commercialization of solar PV technology to date have now become the impetus and reasons for its growth in recent times. The reasons are technological advancements, the reduced cost of PV materials, and government policies. Photovoltaic technology is the indirect but the most efficient way to harness the sun's energy. Due to recent developments, PV manufacturing technology has recorded a worldwide growth of 23.5 GW in 2010 and is expected to further grow at the rate of 35% or more, thus making it the fastest growing industry.

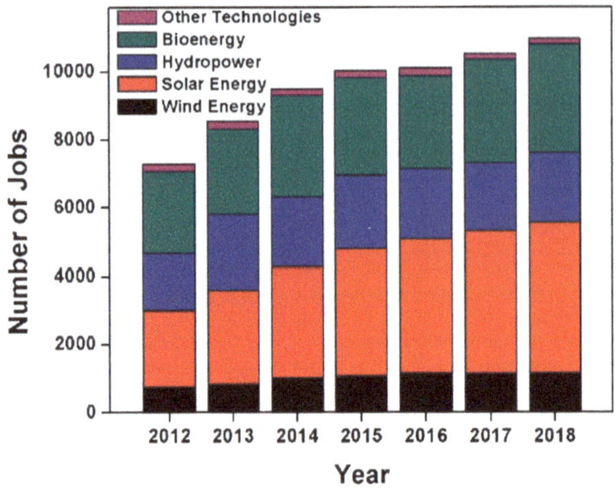

Figure 1.3 Fast-growing employment prospects in the renewable energy sector. *(Source: IRENA jobs database).*

Researchers are continuously striving to improve the power conversion efficiency of solar cells that is critical for the growth of the PV industry [11].

A solar PV cell is an electronic device that directly converts sunlight into electricity through a photovoltaic process, which is similar to the process of "photosynthesis" in plants. The phenomena of photovoltaics happens due the absorption of photons of sunlight inside the absorber layer, which is nothing but a highly crystalline direct-bandgap semiconductor thin film. The series-parallel connection of cells constitutes a PV module, which also includes various encapsulation materials to protect the module from external factors like oxygen, dust, and humidity. The details of the solar PV cell structure will be described in detail in the subsequent chapter. Some popular examples of solar cells are thin-film solar cells, organic solar cells, dye-sensitized solar cells, and perovskite solar cells, among others.

The main difference between the aforementioned cells of succeeding generations is the type of absorber layer used in their manufacturing. A few examples of the absorber layers used are monocrystalline silicon (mc-Si), gallium arsenide (GaAs), cadmium telluride (CdTe), copper indium gallium diselenide (CIGS), polymer, perovskite etc. [12,13], and some low-dimensional materials like carbon nanotubes and quantum dots. During the initial days of PV manufacturing, crystalline silicon was the most preferred material, but it was replaced by CdTe which has a more suitable bandgap, although soon it became less popular due to the involvement of toxic elements. CIGS material was less toxic, as well as having the capability for bandgap tunability, making its case very strong compared to CdTe materials. These absorber layers have the drawback that they need to be deposited at high temperatures under vacuum conditions, which makes the PV manufacturing highly expensive.

As an alternative, using processed organic materials provides the lucrative benefit of thin-film deposition under room-temperature conditions. However, the crystallinity of these materials is comparatively low, thus producing an inferior power-conversion efficiency in the solar cells and modules. Apart from the absorber layer, several other layers, such as the widow layer, buffer layer, and electron and hole transport layers, also play a critical role in the operation of solar cells. Another important component of PV manufacturing is the substrate material, which can be rigid like glass and silicon or flexible like paper, plastic, and fabric. Depending on the nature of substrate, the PV modules can be either rigid or flexible in nature. Apart from the PV cell, the other aspects that ultimately provide electricity are also important to identify.

The last stage of PV manufacturing, and one of the most critical in the industry, incorporates a suitable encapsulation material to protect the solar cells from dust, moisture, snow, rain, and mechanical vibrations. The encapsulation involves the interconnection of solar cells by soldering in series or parallel arrangements, as per the requirement of power (current and voltage) to feed the load and, at the same time, protect the module from corrosion and mechanical vibrations induced during transportation and installation. The PV manufacturing process is a simplistic one because it does not involve any complex tools and instruments. The performance degradation of PV modules, if maintained properly, is quite slow and provides service up to 30 years. The major steps of PV manufacturing process can be summarized as follows:

1. Selection of PV and related materials
2. Arrangement of glass, use of machines for layer deposition, cutting the rolls of encapsulants and backsheets
3. Encapsulation and packaging
4. Testing procedures like Hi-Pot and electroluminescence tests

1.5 Characteristics of PV cell

A solar cell is a p–n junction diode working under solar illumination. The typical structure of a thin-film solar cell is sketched in Fig. 1.4.

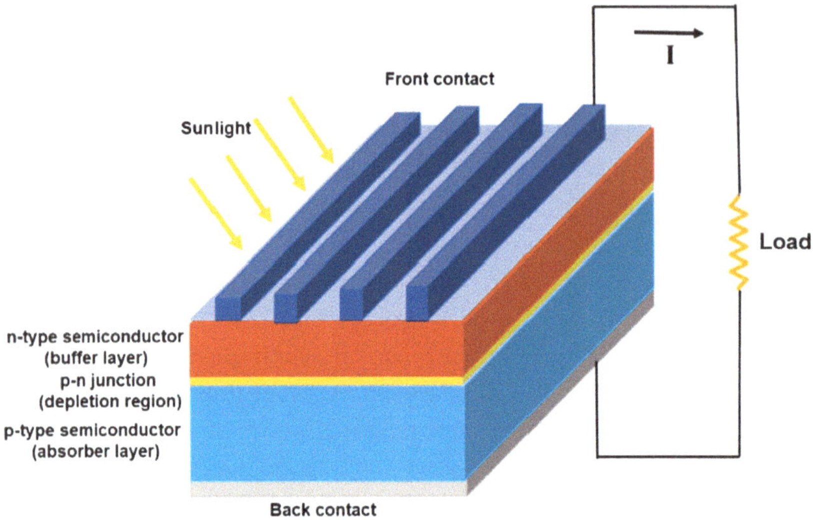

Figure 1.4 Typical solar-cell structure feeding a resistive load.

The depletion region at the interface of p- and n-type materials facilitates the dissociation of excitons (generated inside the absorber layer) into free-charge carriers. Just like any other electronic device, electrical characterization of solar cells is done post-fabrication in terms of various characteristics like current–voltage (I-V) and quantum efficiency–wavelength (QE-λ) [14]. Off these, I-V characteristics yield the maximum information about a solar cell's performance in terms of the parameters, such as the fill factor, power-conversion efficiency, open-circuit voltage, and short-circuit current. In the presence of sunlight, the solar panel charges the battery and also supplies the power to the load directly. When there is no sunlight, the charged battery supplies the required power to the load.

In this subsection, we briefly address the characteristics of solar cells.

1.5.1 Quantum efficiency versus wavelength characteristics

Quantum efficiency is defined as the ratio of the number of charge carriers collected by the terminals of a solar cell to the number of photons of a given wavelength incident upon it. Basically, it is the measure of light-to-electricity conversion ability of a solar cell. The 'QE-λ' measurement is a useful tool for the characterization of any new cell fabricated with novel structures and materials. The better the 'QE-λ' spectral response, the better is the solar cell at converting light into electricity. An example of a typical 'QE-λ' characteristics of a solar cell is shown in Fig. 1.5.

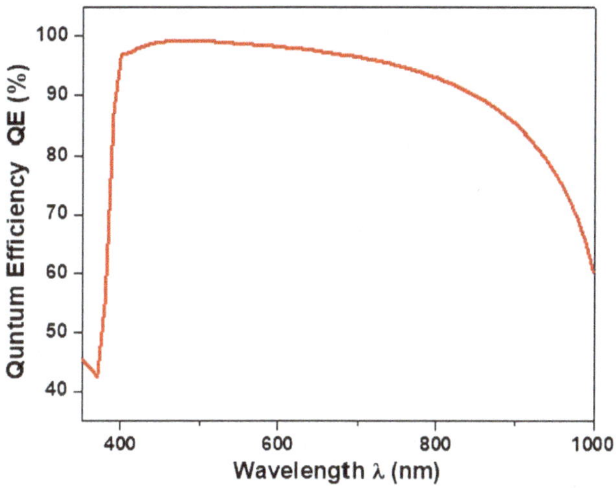

Figure 1.5 Typical 'QE-λ' curve of a solar cell.

1.5.2 Current versus voltage characteristics

The measurement of the I-V characteristics of a manufactured solar cell is the most critical characterization, and it yields a wealth of information related to its performance. When a solar cell is illuminated by sunlight (experimentally done through solar simulator), a photocurrent is generated, flowing in the opposite direction to the forward current under dark conditions. Therefore, the diode current under illumination is given by

$$I_L = I_0 \exp\left(\frac{eV}{nkT}\right) - I_{SC}, \tag{1.1}$$

where I_{SC} is the short-circuit current generated by a solar cell under illumination. A typical I-V characteristics is shown in Fig. 1.6.

The I-V characteristics is used to quantify the solar cell's performance in terms of four parameters:

1. Open-circuit voltage (V_{OC}): It is the voltage measured across the terminals of a solar cell when the external circuit is in an open condition (open terminals) and there is no current flowing through the load connected across the terminals.

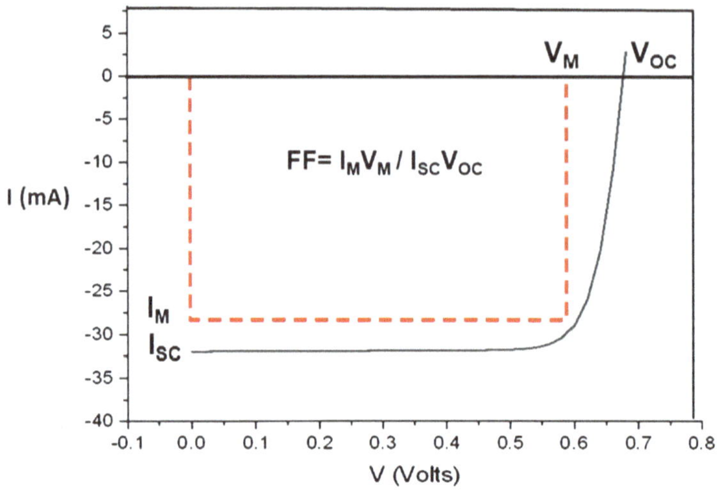

Figure 1.6 Typical I-V characteristics of a solar cell.

2. Short-circuit current (I_{SC}): When the terminals of a solar cell are in a short-circuit condition and the potential difference across the terminals is zero, the current produced is called the short-circuit current. Although, I_{SC} is used with the assumption that the illuminated surface area (contact area) of a solar cell is unity, it is always better to measure the short-circuit current density (J_{SC}) instead of the short-circuit current for the sake of performance comparisons between cells with varying contact areas.
3. Fill factor (FF): This is another important factor of the merit of solar cells; it indicates the deviation of the I-V curve (IVth quadrant) from a perfect square. It is a true measure of the power that can be extracted from the solar cell. Mathematically, it is defined as

$$FF = \frac{V_M I_M}{V_{OC} I_{SC}}, \tag{1.2}$$

where V_M and I_M are the maximum power-extraction points.
4. Power-conversion efficiency η: The sunlight-to-electricity conversion efficiency of a solar cell is called its power-conversion efficiency and is mathematically expressed as

$$\eta = \frac{Output}{Input} = \frac{V_M\, I_M}{P_{in}}, \tag{1.3}$$

where P_{in} is the input power.

The properties of the absorbing material primarily determine the performance of solar cells. However, the cell architecture also affects the cell characteristics. Silicone solar cells comprise 95% of the solar-cell market today. The silicone industry has achieved remarkable progress in minimizing the device's energy losses and maximizing the extraction of the photocurrent. The basic structure of a conventional cell is the so-called Back Surface Field (BSF) cell. BSF cells took 70% of the market in 2017. However, we expect the Passive Emitter Rear Contact (PERC) structure to dominate soon. In this case, a thin layer of aluminum oxide gets added to reduce the carrier recombination at the back surface. It's called the passivation layer. The module consists of connecting the cells and placing them in a sandwich between two layers of the encapsulant. In a traditional silicon cell-based photovoltaic module, the ribbons interconnecting neighboring cells can cause a significant loss of power due to the current transport. Cutting solar cells in half is an effective way to lower resistive power losses, especially in sunny regions. The half-cut cell generates half the current of a standard cell, reducing resistive losses by 75% in the solar modules' interconnections.

1.6 Current status of PV technology

The current decade has witnessed significant developments in PV technology as the PV system installation has increased to more than 100 GW in 2018. The major reason behind such a stupendous rise is the reduction of PV-system prices leading to higher annual installations worldwide. However, a few challenges remain to further enhance the commercial aspect of PV technology. Presently, the European Union is the second largest manufacturer of PV modules after China. But most of the countries are striving hard to get a foothold in PV installation capacity because this will reduce their large sums of GDP spent for purchasing crude oil. The major hurdle for PV technology is its associated economic burden and the strong measures required to be taken by governments to make this technology cost effective.

Already some major developments have happened in PV technology that auger well for its future prospects. The prices of PV panels have drastically dropped by 20%, with a doubling of manufacturing volume driven by technological advancement and market growth. The market price of PV modules around the globe have plummeted in the last decade owing to the reduced dependence on silicon which is used as the raw material for their manufacturing. Also, a decade earlier there had been a shortage of the material that prevented lower pricing. But, with the rise of thin-film solar-cell technology- this dependency has been reduced, and it also spawned huge investments from major players in the market. However, the prices of PV installation still vary significantly around the globe because they are largely impacted by the prices of accessories involved such as cables and inverters and also due to the geographical location of various countries and taxes levied by their governments.

The cost effectiveness of power-generation capacity of PV systems is measured using the concept of Levelized Cost of Energy (LCOE), which is the price at which electricity should be generated from a source and is the economic assessment of the system, which includes its service time, capital investment, operation, and maintenance. The future growth prospects of PVs are continuously being explored by the International Renewable Energy Agency (IRENA) in two major aspects. The first one is through technology-friendly economic policies, while second aspect is the realization of technology with cleaner and climate-favorable alternatives. Some major points of IRENA's renewable-energy road map are:

1. Identifying the long-term plans and government policies required for the global expansion of technology meeting the Paris Agreement.
2. To work towards the optimization of low-carbon technologies to make the PV technology more energy and environment friendly. Apart from this, more effective approaches for electrification are also required.
3. A more realistic and holistic energy-transformation scenario called the 'REmap Case' need to be developed that can more efficiently deal with the rapid deployment of low-carbon technologies.
4. The better projection and anticipation of investment needs and cost benefits associated with the future low-carbon technologies to achieve the energy transformation.

Presently, while the PV systems as a whole have shown tremendous growth prospects over the years, there still are some major path-breaking developments required to make this technology affordable by the masses. In this context, some major issues need to be seriously dealt with, especially at the cell level, which primarily include the phenomena of potential induced degradation (PID) and the soiling effect coupled with the reduction in necessary cleaning. These two factors are largely responsible for increasing the failure rate of PV installations. The PID and soiling effects come into play after the successful installation of PV systems, and special measures are required to be taken both at the cell and the system levels to eliminate these. Some key aspects of future PV growth are as follows:

1. Better optimization of PID effect to enhance the longevity of the installation.
2. Identification of more sustainable and cost-effective alternatives to the currently dominant silicon and compound semiconductor materials used for the fabrication of solar cells.
3. Improvement of the quality of lamination and flexibility in PV modules required for the enhanced trapping of solar energy.
4. Prevention of performance-degradation phenomena, such as cell cracking and corrosion by adopting innovative approaches.
5. Better maintenance techniques to remove/reduce the problem of frequent dust accumulation (soiling effect) and biological soiling of PV panels.
6. Requirements for highly skilled personnel for data acquisition, geographical surveying, long-term monitoring of outdoor conditions, and PV system-failure surveying.

Future PV technologies are bound to face several major obstacles on the path to replacing conventional fossil-fuel energy-generation systems. Some useful strategies that can be adopted for a sustained growth of PV technology are listed here:

i Ensuring drastic reduction in off-grid products
ii Government subsidies to establish grid-connected solar parks
iii Promoting solar rooftop installations to reduce the need for establishing solar parks
iv Encouraging the building-material suppliers to assist in the growth of PV technology
v Exploring the development prospects of innovative concepts like bio-solar cells, floating solar farms, wireless transfer of solar power from space, and solar energy-harvesting trees
vi Developing extensive strategies to improve the power-conversion efficiency, reliability, and the manufacturing ease of solar cells
vii Emphasizing the expansion of the International Solar Alliance.

1.7 Does PV pay back?

People are curious about photovoltaics because they understand that it takes a certain amount of energy to create a photovoltaic module and there are unavoidable emissions tied to creating those modules. There are undoubtedly financial obligations connected to making and purchasing and installing a photovoltaic module. So we ask ourselves, or address the question in general to people who install them, when does PV payback? We can divide the response into three available frameworks. The first is an energetic payback. The second is when a financial payback will happen for the client, and the third would be the onset of emissions payback. Part of the decision-making for installing photovoltaic panels relies on the idea that solar-to-electric conversion would be a carbon-neutral strategy or a negative-carbon strategy for photovoltaic technology. In essence, it replaces other technologies, such as coal combustion that otherwise would emit carbon dioxide gases into the atmosphere, if we didn't have photovoltaics.

Probably one of the key factors in the decision to invest in a photovoltaic module and an array of photovoltaics on your residential building, commercial building, or a solar farm is going to be the financial payback. So, when does this investment return and pay itself back financially? It's a factor of both the sunlight conditions, the quality of sunlight, the intensity of sunlight in the location you are living, and the alternative energy cost you want to replace. Despite a powerful solar resource, photovoltaics don't seem to be paying themselves back as fast in some places. The other payback is the energetic payback. A particular amount of energy gets used to build the photovoltaic modules comprised of solar cells. That high-energy density to make the technology has to be replaced. So, when we deploy a photovoltaic module, it starts creating energy, and it's going to replace that energy ultimately invested in manufacturing it [15,16]. The time required to pay that back is between two to five years, depending on the geographical location.

The third form of payback refers to emissions and greenhouse-gas emissions, so manufacturing a photovoltaic module requires energy. That energy is very likely to be sourced from coal combustion. The manufacturing step will embed a lot of energy and with it a lot of CO_2. The resulting photovoltaic module, as a result of being deployed in the sun, will produce electricity that is carbon-free. The amount of negative emissions or generating electricity without carbon-dioxide emissions will ultimately pay itself back in a few years. There are recommendations in the literature to reduce such payback periods [17,18].

In the next few chapters, a comprehensive discussion of the issues related to the development of PV technology and possible solutions are presented. In total, the chapters are intended to give readers a holistic understanding of the concept of PV and various aspects related to the development of cells and modules, along with some nontechnical ones for the sustainable growth of PV technology.

The next chapter brings together a detailed discussion pertaining to some important features of thin-film photovoltaics. The discussion is expected to provide the readers an exhaustive but concise overview of the concepts related to the photovoltaic energy-conversion process. Also, we provide a brief introduction to the conventional and emerging materials that are required for fabrication of thin-film photovoltaic cells and modules. The following chapter should make the reader comfortable in understanding the gruelling concepts of performance degradation, mitigation, and forecasting approaches required to adopt thin-film photovoltaics.

Bibliography

[1] I. Overland, Energy: the missing link in globalization, Energy Research & Social Science 14 (2016) 122–130, https://doi.org/10.1016/j.erss.2016.01.009.

[2] S. Sorrell, Reducing energy demand: a review of issues, challenges and approaches, Renewable & Sustainable Energy Reviews 47 (2015) 74–82, https://doi.org/10.1016/j.rser.2015.03.002.

[3] G. Singh, Solar power generation by PV (photovoltaic) technology: a review, Energy 53 (2013) 1–13, https://doi.org/10.1016/j.energy.2013.02.057.

[4] M. Sharmina, C. McGlade, P. Gilbert, A. Larkin, Global energy scenarios and their implications for future shipped trade, Marine Policy 84 (2017) 12–21, https://doi.org/10.1016/j.marpol.2017.06.025.

[5] R. Li, M. Su, The role of natural gas and renewable energy in curbing carbon emission: case study of the United States, Sustainability 9 (4) (2017) 600, https://doi.org/10.3390/su9040600.

[6] L. Al-Ghussain, Global warming: review on driving forces and mitigation, Environmental Progress & Sustainable Energy 38 (1) (2018) 13–21, https://doi.org/10.1002/ep.13041.

[7] D. Deb, Intelligent decision making device for residue incorporation in soil or biomass power plants, Journal of Intelligent & Fuzzy Systems 38 (5) (2020) 6065–6074, https://doi.org/10.3233/JIFS-179691.

[8] H.S. Dhiman, D. Deb, Decision and Control in Hybrid Wind Farms, Springer, Singapore, 2020.

[9] S. Mishra, K. Bhargava, D. Deb, Numerical simulation of potential induced degradation (PID) in different thin-film solar cells using SCAPS-1d, Solar Energy 188 (2019) 353–360, https://doi.org/10.1016/j.solener.2019.05.077.

[10] D. Deb, N.L. Brahmbhatt, Review of yield increase of solar panels through soiling prevention, and a proposed water-free automated cleaning solution, Renewable & Sustainable Energy Reviews 82 (2018) 3306–3313, https://doi.org/10.1016/j.rser.2017.10.014.

[11] B. Parida, S. Iniyan, R. Goic, A review of solar photovoltaic technologies, Renewable & Sustainable Energy Reviews 15 (3) (2011) 1625–1636, https://doi.org/10.1016/j.rser.2010.11.032.

[12] A.E. Amrani, A. Mahrane, F.Y. Moussa, Y. Boukennous, Solar module fabrication, International Journal of Photoenergy 2007 (2007) 1–5, https://doi.org/10.1155/2007/27610.

[13] V. Tyagi, N.A. Rahim, N. Rahim, J.A. Selvaraj, Progress in solar PV technology: research and achievement, Renewable & Sustainable Energy Reviews 20 (2013) 443–461, https://doi.org/10.1016/j.rser.2012.09.028.
[14] L.L. Kazmerski, Photovoltaics: a review of cell and module technologies, Renewable & Sustainable Energy Reviews 1 (1–2) (1997) 71–170, https://doi.org/10.1016/s1364-0321(97)00002-6.
[15] D. Kapoor, P. Sodhi, D. Deb, Solar panel simulation using adaptive control, in: 2012 IEEE International Conference on Control Applications, IEEE, 2012.
[16] D. Kapoor, P. Sodhi, D. Deb, A novel control strategy to simulate solar panels, in: 2012 International Conference on Signal Processing and Communications (SPCOM), IEEE, 2012.
[17] D. Wijesuriya, K. Wickramathilaka, L. Wijesinghe, D. Vithana, H.R. Perera, Reduction of solar PV payback period using optimally placed reflectors, Energy Procedia 134 (2017) 480–489, https://doi.org/10.1016/j.egypro.2017.09.606.
[18] I. Celik, A.B. Philips, Z. Song, Y. Yan, R.J. Ellingson, M.J. Heben, D. Apul, Energy payback time (EPBT) and energy return on energy invested (EROI) of perovskite tandem photovoltaic solar cells, IEEE Journal of Photovoltaics 8 (1) (2018) 305–309, https://doi.org/10.1109/jphotov.2017.2768961.

CHAPTER 2

Thin-film photovoltaics

In this chapter, we provide a detailed discussion covering the essential aspects of thin-film photovoltaics (TFPV). A photovoltaic device consists of several vital layers, and it is mandatory to gain insight into their utility to achieve further improvement in photovoltaic performance. Moreover, to understand the concepts related to performance degradation in thin-film photovoltaics, it becomes necessary to get acquainted with the basic photovoltaic energy-conversion principle. Since the most crucial layer inside a thin-film photovoltaic cell is the absorber layer, therefore, this chapter presents an introduction to different generations of absorber layer materials.

2.1 Introduction

Due to the growing concern of global warming, the demand for renewable energy sources has seen astonishing progress like never before. Among several available renewable energy sources, the research into and development of photovoltaic (PV) technology are progressing with each passing day [1]. PV technology's importance lies in the fact that it is one of the cleanest technologies available to date. It directly converts solar energy to electricity without yielding any nonecofriendly byproducts. PV technology development requires a multidisciplinary approach because it involves concepts related to electronics, physics, chemistry, and material science. Although the PV industry has recently grown remarkably, some significant challenges are still there, such as the availability of inexpensive and nontoxic raw materials, a simple manufacturing process flow, high-power conversion efficiency, flexibility, and improved durability [2]. We expect an ideal PV cell to completely absorb the solar spectrum (Fig. 2.1) incident on it, which depends entirely on the absorber layer's properties inside the cell.

Usually, the sunlight reaching Earth's surface (terrestrial radiation) has a significantly lower intensity than the extraterrestrial radiation attributed to absorption while passing through the atmosphere. The solar spectrum reaching the Earth's surface is known as AM1.5, representing the air mass corresponding to a zenith angle of 48.2°. The average intensity of this radiation is 1000 mW/cm^2 (1 sun) [3]. The solar spectrum shown in Fig. 2.1 demonstrates the huge amount of energy received by the Earth's surface,

Degradation, Mitigation, and Forecasting Approaches in Thin Film Photovoltaics
https://doi.org/10.1016/B978-0-12-823483-9.00013-9

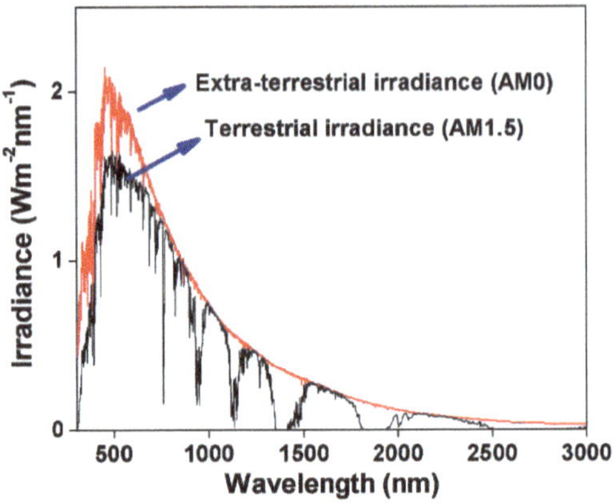

Figure 2.1 Comparison of the extraterrestrial (AM0) and terrestrial (AM1.5) solar-irradiance spectra.

which humanity has still not utilized to its full potential. We can harness the solar energy in two ways, namely, 'solar-thermal technology' (utilizes the infrared portion of the spectrum in the form of heat energy) and 'photovoltaic' (utilizes mainly the visible portion of the spectrum) [4]. Therefore, we dedicate this book to the studies on PV technology at the cellular and modular levels.

A TFPV cell is composed of thin layers of conductors, semiconductors, and insulators. The configuration of a TFPV cell and its performance are closely related to each other. The most crucial layer inside a TFPV cell is the absorber layer, which is generally a p-type semiconductor material with a high absorption coefficient. The presence of the p–n junction inside these cells is responsible for generating the electron-hole pairs upon absorption of photons coming in the form of sunlight. This chapter presents a brief overview of the terms related to TFPV operation and manufacturing.

2.2 Photovoltaic effect

The operation of TFPVs is based on the photovoltaic effect, which refers to the generation of voltage/current inside the cell when sunlight falls on it and subsequently gets absorbed in the absorber layer to produce free carriers [5]. A TFPV cell is an electronic device that generates free charge carriers due to the photovoltaic effect. A p–n junction inside the cell induces a built-in potential that facilitates exciton dissociation into free charge carriers. The harnessing of solar energy through the PV technique is described in Fig. 2.2.

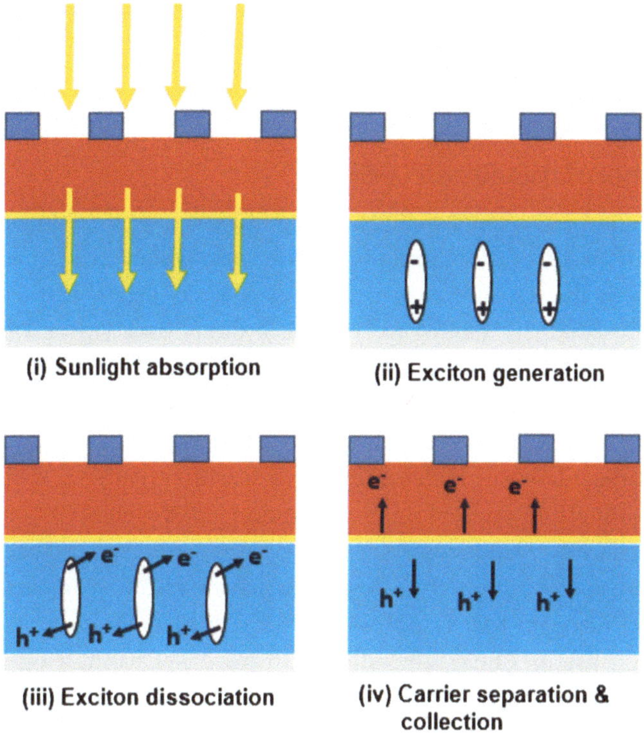

Figure 2.2 Process of photovoltaic energy conversion in TFPVs.

The photovoltaic effect's efficacy is governed by competing phenomena of carrier generation, recombination, and transport. This technique involves the following four-step sequential process:

1. Absorption of incident photons inside the absorber layer
2. Generation of excitons (bound electron and hole pair)
3. Exciton dissociation into free electrons and holes
4. Separation and collection of free electrons and holes at the respective electrodes (generation of electricity).

2.3 Some popular TFPV absorber materials

TFPVs are generally categorized based on the absorber layer material appropriate for usage in their fabrication. From the commercialization aspect, the essential properties of absorber materials are a low processing cost, high optical absorption coefficient, direct and narrow band gap, high carrier mobility, useful carrier lifetime, low defect density, and superior thermal stability [6]. The emergence of TFPVs is based on a variety of absorber materials that are used in their manufacturing. Such details are briefly described as follows.

2.3.1 Amorphous silicon (a-Si)

Crystalline silicon consists of a well-ordered lattice, while amorphous silicon (a-Si) consists of a continuous random network which causes a-Si to have a much higher defect density. These defects create allowed energy states in the bandgap. Therefore, a-Si has a direct bandgap of 1.6–1.8 eV, while crystalline silicon has an indirect bandgap of 1.1 eV. The direct bandgap of a-Si leads to higher absorption, by about an order of magnitude, in the spectrum's visible and ultraviolet parts.

a-Si is regarded as a high-quality absorber-layer material owing to its lucrative properties, such as high absorption coefficient (10^4 cm^{-1}), direct bandgap, low manufacturing cost, and shorter energy payback period [7]. However, a-Si suffers from low crystallinity, resulting in sub-standard carrier mobility and carrier diffusion length and large defect density that restrict their use in TFPV fabrication. On the other hand, it offers significant efficiency and scope for performance improvement, making it one of the most pursued absorber-layer materials for TFPV fabrication. The major limitation of the a-Si layer arises from dangling bonds present inside the crystal structure that cause shorter carrier diffusion length, which can be improved drastically through hydrogen passivation (a-Si:H) [8]. However, a-Si:H-based TFPVs tend to undergo light-induced degradation (LID), thus limiting their power-conversion efficiency (1% per year) [9]. Even though a-Si:H is a disordered material, its ability to undergo doping makes it a suitable candidate for PV applications.

Another interesting fact about the a-Si:H absorber layer is its high defect density and inferior carrier mobility inside the crystal structure. Therefore, a-Si:H-based TFPVs are generally fabricated in either p–i–n (superstrate) or n–i–p (substrate) configurations. In both p–i–n and n–i–p configurations, the sunlight strikes from the p-side to generate the electron-hole pairs near the p/i interface. Some popular techniques used to deposit a-Si's films:H are chemical vapor deposition (CVD), sputtering, ion-beam assisted evaporation, and expanding thermal plasma deposition. The plasma-enhanced chemical vapor deposition (PECVD) technique is mostly used [10]. In a-Si:H materials, both band-tail and mid-gap defect states create the most favorable recombination sites, reducing the efficiency of based TFPV cells. Between these two, the mid-gap defect states play a more dominant and irreversible effect on charge carrier trapping, leading to enhanced recombination of photo-generated electrons and holes. Several characterization methods are generally adopted to predict the suitability of an a-Si:H layer for TFPV fabrication, e.g., spectroscopic ellipsometry, constant photocurrent method, electron spin resonance (ESR), time-of-flight (ToF), and hydrogen effusion [11].

2.3.2 Cadmium telluride (CdTe)

A CdTe TFPV is an excellent alternative to the expensive silicon wafer-based PV technology. The properties of CdTe materials, such as a high absorption coefficient and a direct bandgap (1.45 eV), augers well for the industrial manufacturing of low-cost PV modules [12]. The current market capture of a-Si:H-based TFPVs is due more to their superior efficiency, but this technology's cost is still very high due to an expensive fabrication process. The CdTe technology offers a more sustainable and inexpensive (shorter energy-payback time) alternative. The credit goes to the high value of its absorption coefficient resulting in reduced material wastage for mass manufacturing of TFPV modules [13].

Moreover, its stability at high processing temperatures has quickly made it a lucrative alternative to a-Si:H-based TFPVs. A typical CdTe cell involves an n-type buffer layer (mostly CdS) to form a hetero-junction that assists in dissociating excitons into free charge carriers [14]. One of the major performance-limiting issues in CdTe TFPVs is the diffusion of metal atoms during the deposition of back contact producing recombination sites inside the absorber layer [15]. This technology's major drawback is the toxic nature of cadmium. The scarcity of the other main component in CdTe, tellurium, is also a limitation.

A vital strategy generally employed to improve CdTe TFPVs is the absorber layer's heat treatment irrespective of its deposition method. The improvement in post-heat treatment performance is attributed to the large grain size and defect passivation (carrier-lifetime improvement). The thin-film properties of the CdTe layer complements well with both rigid and flexible substrates. The frequently used CdTe thin-film deposition techniques are close-spaced vapor transport (CSVT), sputtering, laser-ablation spray pyrolysis, and electrodeposition [16].

Next, we study the current market share of cadmium telluride solar cells and how the conversion efficiency has developed over time. The most significant market share comes from First Solar, a US company with production facilities in the US, Malaysia, and Germany. The annual production in 2015 was about a 2.5 giga-watt-peak, which is large compared to CIGS and amorphous silicon, but small compared to crystalline silicon with about a 57 giga-watt-peak in 2015. Together with cadmium, it forms a two–six semiconductor compound. Cadmium is a so-called transition metal, meaning that it has a partially filled d electron shell. Cadmium telluride has a bandgap of 1.44 eV, which, as we can see, lies very close to the optimum value for the bandgap of a single-junction solar cell.

A layer of a few μm is enough to absorb all photons with an energy above the bandgap. Each atom of cadmium has tetrahedral bonds with four neighboring tellurium atoms. At room temperature, the lattice constant is 0.648 nm. Cadmium-telluride material can be grown as both p- and n-types. It is, however, very problematic to achieve moderate to high doping concentrations in cadmium telluride. These elements have one more valence electron than cadmium, which will contribute to the free-electron concentration. On the other hand, n-type doping is achievable by replacing tellurium atoms with group-seven elements, such as fluorine, chlorine, bromine, and iodine. Finally, intrinsic defects in the material also contribute to the conductivity type. For instance, a missing tellurium atom (a tellurium vacancy) will result in more free electrons in the material since a tellurium atom would accept two free electrons.

Copper is a very frequently used acceptor element in cadmium-telluride solar cells. Still, it is essential to limit the diffusion of copper into other device layers because this will degrade the cell's performance. Acceptor atoms on a tellurium site can be, for instance, phosphorus, arsenic, or antimony. Finally, a cadmium vacancy will also act as an acceptor-type defect because, if this cadmium atom were present, it would contribute two free electrons. In the standard cadmium-telluride cell design, the p-type material is the absorber layer. The material is challenging to dope, but shallow doping is possible with substitution doping and intrinsic defects. The market leader First Solar has been expanding its production capacity over the last couple of years. At the same time, the conversion efficiency of lab-scale solar cells is very promising.

2.3.3 Copper indium gallium selenide (CIGS)

Next, we introduce the group of chalcopyrite solar cells known as the copper indium gallium diselenide (CIGS). Both cadmium telluride and CIGS are thin-film solar-cell technologies. The National Renewable Energy Laboratory, NREL, in the US keeps track of the record conversion efficiencies for all solar-cell technologies. The record efficiency for a CIGS cell under 1-sun illumination is 22.6%, implying a high potential for this technology. Chalcopyrite materials consist of elements in groups one, three, and six. Many combinations of these elements are potential solar-cell materials. The electronic bandgap of most materials is too wide. The most common combination is a mixture of copper indium diselenide or CIS and copper gallium diselenide or CGS. Also, sulfur can be included in the structure, partially replacing the selenide fraction in the material. However, the iron atoms are replaced by gallium and indium atoms, while sulfur is wholly or partially replaced by selenium.

The band gap varies over 1–1.7 eV by changing the ratio between the pairs indium and gallium and sulfur and selenide. Adding gallium to the structure, for instance, increases the bandgap of the material. This material has a very high absorption coefficient, which makes it a suitable material for thin-film technology. On top of this, the diffusion length of electrons can be up to a few micrometers. We know that a p- or n-type material can be obtained by adding impurity atoms to the structure. The type of conductivity in CIGS is determined by the material's intrinsic chemical composition, and p- or n-type conductivity can be obtained by changing the material composition in two different ways. In the standard CIGS technology, p-type material is used as an absorber layer. CIGS's conductivity type is determined by the material's composition rather than the extrinsic doping of impurity atoms.

CIGS absorber material has been receiving worldwide attention for TFPV manufacturing owing to its great potential to replace the crystalline silicon-wafer technology since it can be deposited over the low-cost rigid (e.g., glass) and flexible substrates (e.g., polyethylene terephthalate) with comparable efficiency value (>22%) [17]. However, the increased film-deposition complexity presents a vast scope for future research in this area. CIGS offers distinctive absorber-layer properties like a direct bandgap and high absorption coefficient (10^5 cm^{-1}). Moreover, its bandgap can be easily tuned through compositional variation, and also the material wastage is drastically reduced, leading to an enormous reduction in the manufacturing cost of PV modules. The inclusion of Ga atoms is deliberately done to tune the bandgap to achieve better matching with the solar spectrum. Another essential feature of the CIGS layer is that it offers both terrestrial and space applications due to its radiation-tolerance capability [18].

Apart from the advantages just mentioned, CIGS materials present several commercialization limitations. The limited availability of expensive indium and compositional variations inside thin films, ranging from run to run and causing large variations in efficiency, are significant obstacles in commercializing CIGS TFPVs. Also, the primary cell-level performance limitation arises from the presence of mid-gap defect states attributed to the dangling bonds present inside the thin film, which lead to enhanced recombination and interfacial discontinuity of energy levels.

The most effective strategy used to counter the CIGS layer's limitations is bandgap grading, which reduces recombination losses [19]. Some popular CIGS growth techniques are co-evaporation, sequential sulfurization, sputtering, and other nonvacuum techniques like electrodeposition, spray coating, screen printing, and spin coating. The nonvacuum methods are beneficial in reducing material wastage and cost, superior stoichiometric control, and roll-to-roll fabrication compatibility.

2.3.4 Copper zinc tin sulphide (CZTS)

CZTS is the next most promising, nontoxic, low-cost, and eco-friendly inorganic absorber layer material used in TFPVs. A thin film of CZTS appears as a black crystal with a greenish tinge. Like CIGS, it also has good sunlight absorbing and electronic properties, thus attracting significant PV research interest worldwide. The easy availability of composite elements on the Earth's crust makes the deposition of CZTS thin films quite economical [20]. On the contrary, the deposition of CIGS thin films are costly due to the requirement of using rare and expensive elements like indium and gallium.

Moreover, the appropriate bandgap value (1.4–1.5 eV) and high-temperature stability (1200K) of CZTS material makes it a highly lucrative alternative for TFPV fabrication [21]. The popular synthesis methods for CZTS thin films are electrochemical deposition, vacuum deposition, electron-beam evaporation, magnetron sputtering, spray pyrolysis, pulsed-laser deposition, and the sol-gel process. For thin films of CZTS layers, some useful characterization techniques are photoluminescence spectroscopy, Raman spectroscopy, X-ray diffraction, admittance spectroscopy, and drive-level capacitance profiling. A critical drawback of CZTS material is that they undergo phase transitions leading to defect states inside the crystal structure [22].

2.3.5 Organic semiconductors

The carbon-based compounds and their derivatives are classified as organic semiconductors. Between the two types of organic semiconductors, i.e., small molecules and conducting polymers, the latter is the preferred choice of an organic material for TFPV applications. The reasons are its attractive features like room-temperature-based solution-processability, low cost, high absorption coefficient, tunable bandgap mechanical flexibility, light weight, and biodegradability [23]. Compared to inorganic semiconductor-based TFPVs, the ones fabricated with organic semiconductors suffer from low efficiency, limited carrier lifetime, and ambient stability. The primary factor behind the performance limitation of organic material-based TFPVs is the organic absorber layer's low crystallinity (morphology).

The popular thin-film deposition techniques of organic semiconductors used during the fabrication of organic PVs are casting, spin coating, doctor blade, screen printing, inkjet printing, spray coating, and the roll-to-roll method [24]. The preferred characterization techniques for organic semiconductors are photoluminescence, atomic force microscopy, Raman spectroscopy, and absorption spectroscopy. The organic TFPV configurations rigorously investigated in the past are metal/polymer junction cells, planar cells, dye-sensitized cells, and bulk heterojunction cells. Organic TFPV technology rates above others due to the mechanical flexibility they offer to install PV panels on curvilinear (nonflat) surfaces. It is straightforward to deposit thin films of organic layers on a flexible substrate at room temperature using solution-processing methods.

2.3.6 Organic-inorganic hybrid perovskite semiconductors

Perovskite materials are the latest and most promising materials that have emerged for TFPV applications due to their lucrative optoelectronic properties, such as high carrier mobility, long carrier lifetime, high absorption coefficient, bandgap tunability, and processing ease. The perovskite materials offer the advantages of inorganic and organic materials because they exhibit optoelectronic properties similar to those of inorganic materials. At the same time, their thin films can be deposited generally in the same manner as organic materials. In a short span of only ten years, perovskite-based TFPVs have reached an efficiency value of more than 22%, and such progress makes them the best alternative to a-Si, CdTe, and CIGS TFPVs [25]. The perovskite TFPVs are also fabricated in both n–i–p and p–i–n configurations through easy and inexpensive deposition methods like single-step deposition, two-step deposition, thermal evaporation, vapor-assisted solution processing, pulsed-laser deposition, inkjet printing, drop-casting, doctor blade coating, slot-die coating, and electrospray-assisted deposition [26].

Like the others, the perovskite PV technology also suffers from drawbacks like toxicity and stability, which are the bottlenecks in their commercialization. The primary source of toxicity in these PV modules is the presence of lead inside the perovskite materials. Some serious progress has been made towards the development of lead-free perovskite TFPVs. The second critical limitation, i.e., the stability of perovskite TFPVs, arises from the degradation of perovskite materials under oxygen and moisture, increased temperature, UV exposure, ion migration, additives, and light exposure. Despite several issues, researchers have consistently improved the efficiency of TFPV cells based on the described absorber materials. Table 2.1 summarizes the highest reported efficiency to date for TFPV cells based on various absorber materials.

Table 2.1 Summary of highest reported efficiency values for TFPV cells based on several absorber materials.

TFPV Cell	Laboratory	Efficiency (%)
a-Si	AIST [27]	10.2
CdTe	First Solar [28]	21
CIGS	Solar Frontier [29]	23.35
CZTS	UNSW	10
Organic	BNLMS	16.48
Perovskite	SECE	24.4

The multiple laboratories that have provided the highest efficiencies as mentioned in Table 2.1 include Advanced Industrial Science and Technology (AIST), the University of New South Wales (UNSW), Beijing National Laboratory for Molecular Sciences (BNLMS), and the School of Energy and Chemical Engineering (SECE).

The structure of TFPVs consists of the absorber layer, buffer layer, electron and hole transport layers (ETLs and HTLs), front and back contacts, rigid and flexible substrates, and encapsulation materials, as described next.

2.4 Conventional and emerging PV materials

While conventional absorber materials are crystalline and amorphous silicon, emerging absorber materials such as CdTe, and CIGS, the CZTS, organic, carbon nanotube, quantum dot, and perovskites have the potential to obtain great commercial success [30]. The second most critical layers are buffer, electron transport, and hole transport layers, which can be made from various electron- and hole-rich materials. Cadmium sulfide (CdS) is conventionally used in CdTe and CIGS cells.

However, some emerging transparent conducting oxide (TCO) materials like zinc oxide (ZnO) and tin oxide (SnO_2) have also shown great promise for low-cost TFPV application. Some popular ETLs used in TFPVs are CdS, PCBM, Al_2O_3, and TiO_2, while CuSCN, NiO, Spiro-OMeTAD, PEDOT: PSS, P3HT, and MoO_3 are used as HTL materials. ITO, FTO, and ZnO: Al are commonly used as front electrodes, and Al, Ag, Cu, and Mo are used as back contacts [31]. TFPVs are classified based on substrate material used during manufacturing. Early on, silicon wafer and glass were commonly used as rigid substrates, while ultrathin glass, metal foil, PET, Upilex-S, Kapton HN, and PDMS are some of the conventional and emerging flexible substrates.

A flexible substrate's quality is generally evaluated in terms of the toughness and flexibility that it can offer. Encapsulation is the final step in TFPV fabrication, protecting the thin films from external entities like oxygen, moisture, and dust. Several encapsulation materials are available for use in commercial TFPV manufacturing. Such materials include ethylene-vinyl acetate (EVA), silicone, thermoplastic polyurethane (TPU), polyvinyl butyral (PVB), UV-cured epoxy resin, Ionomer, ethylene methyl acrylate (EMA), polyisobutylene (PIB), cyclized perfluoro-polymer (Cytop TM), organic materials like poly(vinyl phenol) (PVP), poly(methyl methacrylate) (PMMA), and poly (vinyl alcohol) (PVA). Some of the important quality metrics considered while selecting an encapsulant for TFPVs are the oxygen transmission rate (OTR), water vapor transmission rate (WVTR), glass transition temperature (Tg), total light transmission, water absorption, tensile modulus, UV absorption degradation, and hydrolysis [32].

2.5 Photovoltaic energy conversion

The PV energy-conversion process begins at a solid-state TFPV cell (described in Sect. 2.2) and ends at the PV system. The whole PV system does not involve any moving parts, and it operates quietly with high reliability.

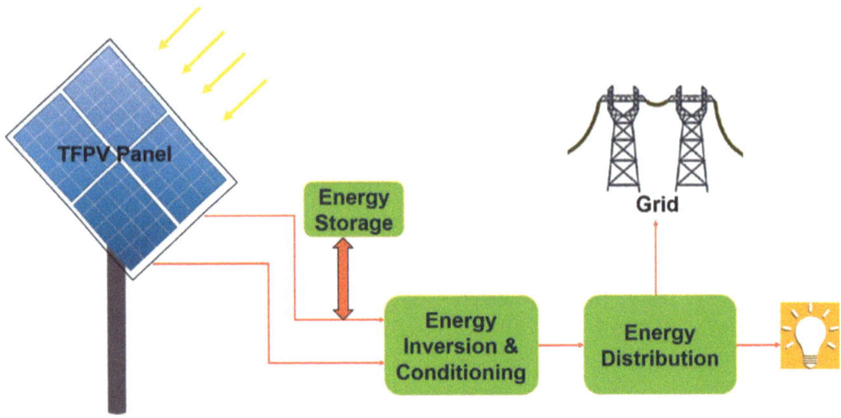

Figure 2.3 Cartoon depicting the principle of PV energy conversion.

Fig. 2.3 depicts the entire process of PV energy conversion, which begins with the PV panel (array of TFPV cells) where sunlight is absorbed and DC power gets generated. Various aspects of energy-storage systems involved in renewable-energy systems have been discussed in the literature [33–35]. We need power conditioning equipment to protect storage batteries, convert DC into AC if the system needs integration with AC equipment, and extract power maximally from solar cells.

The generated electric power can either be stored in a storage device or converted to AC power (energy inversion and conditioning) and distributed to the household utility. Optimally, one directs the remaining electric power to the grid. This whole PV system is known as a "grid-connected PV system" [36]. Its simplicity, installation, and maintenance are relatively easy, and a long service period is also expected. Moreover, one can deploy such systems on transports such as vehicles and spacecraft, although the storage component is still required to be of high capability.

2.6 Thin-film photovoltaic manufacturing and recycling

Generally, the construction of TFPV panels requires high-precision manufacturing and testing equipment. The various components of TFPV panel construction are shown in Fig. 2.4.

TFPV panels consist of the following essential components:

1. TFPV cell matrix: The matrix structure of TFPV cells facilitates the conversion of sunlight to electricity. Each cell acts as an active site

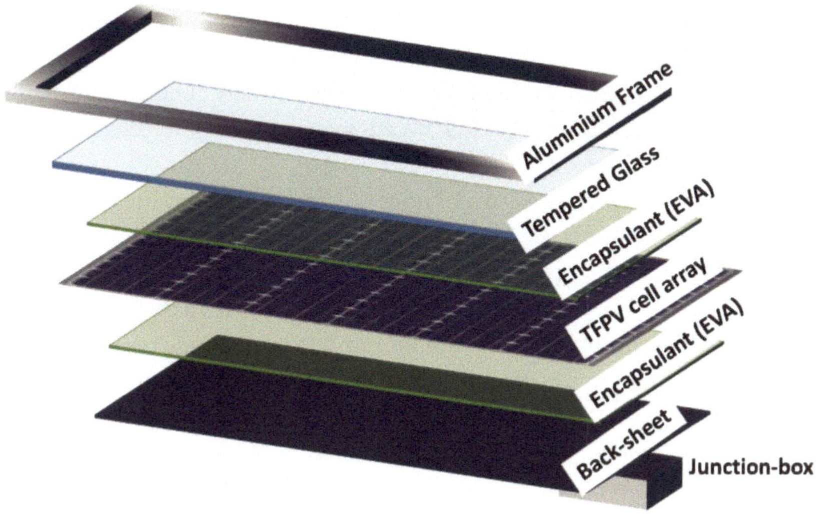

Figure 2.4 Components of TFPV panel construction.

to absorb incident photons inside the absorber layer and generate free electrons and holes. The high crystallinity of the absorber layer helps achieve the maximum power conversion efficiency.

2. Toughened glass: A toughened glass sheet is incorporated to protect TFPV cells from the dust and moisture present in the atmosphere. The typical thickness of a high-strength tempered-glass sheet covering is 3–4 μm and can withstand mechanical vibrations and temperature variations. During the transportation and installation of panels, tempered glass is much better than standard glass. Also, its transparency is much better, resulting in improved efficiency.
3. Aluminum frame: The aluminum frame is used to protect the edge of lamination housing the TFPV cells and provides robustness to the panel during transportation and installation. The extruded aluminum frame of the TFPV panel should be lightweight and bear the extreme stress created by wind and external forces. The frame is generally silver or anodized black, depending on the manufacturer. The frame corners are tightly screwed or clamped together to provide sufficient mechanical strength to the panel.
4. Encapsulation: The frequently used 'ethylene-vinyl-acetate' (EVA), a highly transparent polymer material, encapsulates the TFPV cell-matrix to hold them together during the manufacturing process. EVA is extremely durable and tolerant to extreme weather conditions of

high temperature and humidity. Since the moisture and dust ingression can damage cells' functionality, encapsulation is a mandatory feature for achieving long-term performance of the panels. The encapsulation done on either side of the TFPV cell-matrix makes the cells and their interconnects shock-resistant.

5. Polymer rear back-sheet: A polymer back-sheet material is used as the rearmost layer inside TFPV panels. It prevents moisture ingression, ensures mechanical robustness, and increases electrical insulation. The various back-sheet materials used in TFPV manufacturing include PET and PVF because these offer a high level of protection, thermal stability, and UV resistance. Back-sheet materials are available in a variety of colors.
6. Junction box: The junction box, generally included in the TFPV assembly is located on the panel's rear side. This part of the assembly helps provide secure interconnections between the panels. An essential inclusion inside the junction box is the bypass diodes employed to prevent back-current occurrence due to shading or dust accumulation. In case of failure of bypass diodes, they are replaced easily by removing the cover of the junction box.

Although, TFPV modules serve for a long but finite duration and also need to be disposed of, which creates electronic waste. Therefore, recycling TFPV modules is required to reduce the environmental impact and recover some value from the disposed modules. The currently available techniques for recycling PV modules are capable of recovering only a small portion of materials. Thus, colossal employment and research opportunities are present in developing PV recycling and waste-management techniques. Since most of the PV waste contains heavy and sometimes toxic metals that contribute to environmental degradation, efficient recycling is required to be invented, ultimately increasing the cost of photovoltaic systems.

The purpose of recycling TFPV modules is to prevent ecological degradation and recover precious metals like copper and silver from the disposed modules. Massive efforts are being undertaken worldwide to develop advanced PV recycling methods to protect the environment and enhance value recovery from the disposed modules. Currently, most recycling methods rely on down-cycling processes. Due to the lack of strong regulatory policies in most countries, the recycling of PVs is still inefficient. The TFPV modules are easily recyclable, yielding important constituents like glass, aluminum, and semiconductors that can be quickly recovered and reused. The recycling techniques employed help recover and reuse necessary materials, although some are in the research phase and await commercialization.

A brief summary of commercialized TFPV recycling methods include:

1. Physical disintegration: Provides efficient waste treatment but produces dust particles of heavy metals, causes TFPV cell breakage, and equipment corrosion [37].
2. Thermal treatment: Ensures complete removal of encapsulant, cells remains intact, and constitutes an economical and straightforward technique. Energy consumption is high, harmful emission and cell degradation [38].
3. Dry etching: The most straightforward method but energy requirement is very high, and purification is challenging to achieve [39].
4. Dry and wet mechanical processes: No chemical is involved during processing, the energy requirement is low, and equipment is readily available, but dissolved particles are not removed [40].
5. Chemical etching: One of the most convenient and efficient methods used to recover high-purity materials is chemicals [41].

Apart from the techniques of TFPV recycling just mentioned, there are several emerging techniques as well which are currently in the research phase, and we expect these soon to become an integral part of the TFPV recycling process:

1. Attrition: In this method, chemicals are not required during the recovery of materials which is a utility method for recovering glass. However, chemical treatment is needed [37].
2. Organic solvent dissolution: Minimal cell damage, easy glass recovery but emission of wastes.
3. Mechanical separation by hot-wire cutting: Cell damage is minute, glass gets easily recovered, but the separation of encapsulant requires some other processes [42].
4. Vacuum blasting: This technique is useful for removing the semiconductors and glass without using chemicals, but the recycling method is too slow and causes metal particles' emission [37].

Therefore, both environmental, as well as economic aspects, are associated with the TFPV recycling process. Each recycling method has its pros and cons involved at various different stages, starting from collecting disposed modules to the end of the recycling process. Therefore, to reap the economic benefits, great opportunities are available related to TFPV recycling. We visualize a PV recycling process as the flowchart in Fig. 2.5.

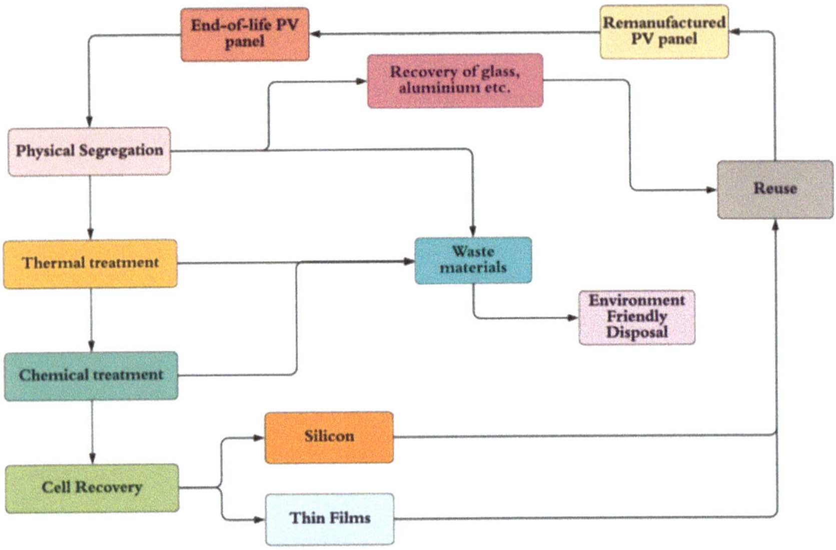

Figure 2.5 Flowchart of PV recycling.

In this chapter, we discussed most of the essential aspects related to TFPV operation technology. Chapter 3 discusses all the critical issues related to performance degradation in TFPV cells and modules, in detail. We expect that such a discussion will yield a clear understanding on the part of readers about the approaches required to mitigate and predict issues related to TFPV systems.

Bibliography

[1] M.Z. Jacobson, Review of solutions to global warming, air pollution, and energy security, Energy & Environmental Science 2 (2) (2009) 148–173, https://doi.org/10.1039/b809990c.

[2] C. Lupangu, R. Bansal, A review of technical issues on the development of solar photovoltaic systems, Renewable & Sustainable Energy Reviews 73 (2017) 950–965, https://doi.org/10.1016/j.rser.2017.02.003.

[3] S.M. Maleki, H. Hizam, C. Gomes, Estimation of hourly, daily and monthly global solar radiation on inclined surfaces: models re-visited, Energies 10 (1) (2017) 134, https://doi.org/10.3390/en10010134.

[4] F. Huide, Z. Xuxin, M. Lei, Z. Tao, W. Qixing, S. Hongyuan, A comparative study on three types of solar utilization technologies for buildings: photovoltaic, solar thermal and hybrid photovoltaic/thermal systems, Energy Conversion and Management 140 (2017) 1–13, https://doi.org/10.1016/j.enconman.2017.02.059.

[5] A.W. Copeland, O.D. Black, A.B. Garrett, The photovoltaic effect, Chemical Reviews 31 (1) (1942) 177–226, https://doi.org/10.1021/cr60098a004.

[6] T.D. Lee, A.U. Ebong, A review of thin film solar cell technologies and challenges, Renewable & Sustainable Energy Reviews 70 (2017) 1286–1297, https://doi.org/10.1016/j.rser.2016.12.028.
[7] B. Rech, H. Wagner, Potential of amorphous silicon for solar cells, Applied Physics A: Materials Science & Processing 69 (2) (1999) 155–167, https://doi.org/10.1007/s003390050986.
[8] M. Stuckelberger, R. Biron, N. Wyrsch, F.-J. Haug, C. Ballif, Review: progress in solar cells from hydrogenated amorphous silicon, Renewable & Sustainable Energy Reviews 76 (2017) 1497–1523, https://doi.org/10.1016/j.rser.2016.11.190.
[9] S.D. Wolf, B. Demaurex, A. Descoeudres, C. Ballif, Very fast light-induced degradation of a-Si:H/c-Si(100) interfaces, Physical Review B 83 (23) (2011), https://doi.org/10.1103/physrevb.83.233301.
[10] A. Terakawa, Review of thin-film silicon deposition techniques for high-efficiency solar cells developed at panasonic/sanyo, Solar Energy Materials and Solar Cells 119 (2013) 204–208, https://doi.org/10.1016/j.solmat.2013.06.044.
[11] Y. Zhao, X. Zhang, B. Yan, Hydrogenated amorphous silicon thin film, in: Handbook of Photovoltaic Silicon, Springer, Berlin, Heidelberg, 2019, pp. 639–692.
[12] B.M. Basol, B. McCandless, Brief review of cadmium telluride-based photovoltaic technologies, Journal of Photonics for Energy 4 (1) (2014) 040996, https://doi.org/10.1117/1.jpe.4.040996.
[13] T. Sinha, D. Lilhare, A. Khare, A review on the improvement in performance of CdTe/CdS thin-film solar cells through optimization of structural parameters, Journal of Materials Science 54 (2019) 12189–12205, https://doi.org/10.1007/s10853-019-03651-0.
[14] S. Mishra, K. Bhargava, D. Deb, Numerical simulation of potential induced degradation (PID) in different thin-film solar cells using SCAPS-1d, Solar Energy 188 (2019) 353–360, https://doi.org/10.1016/j.solener.2019.05.077.
[15] S. Li, Z. Peng, J. Zheng, F. Pan, Optimizing CdTe–metal interfaces for high performance solar cells, Journal of Materials Chemistry A 5 (15) (2017) 7118–7124, https://doi.org/10.1039/c7ta00698e.
[16] L. Kranz, S. Buecheler, A.N. Tiwari, Technological status of CdTe photovoltaics, Solar Energy Materials and Solar Cells 119 (2013) 278–280, https://doi.org/10.1016/j.solmat.2013.08.028.
[17] N. Mufti, T. Amrillah, A. Taufiq, Sunaryono, Aripriharta, M. Diantoro, Zulhadjri, H. Nur, Review of CIGS-based solar cells manufacturing by structural engineering, Solar Energy 207 (2020) 1146–1157, https://doi.org/10.1016/j.solener.2020.07.065.
[18] J. Ramanujam, U.P. Singh, Copper indium gallium selenide based solar cells – a review, Energy & Environmental Science 10 (6) (2017) 1306–1319, https://doi.org/10.1039/c7ee00826k.
[19] F. Ahmad, T.H. Anderson, P.B. Monk, A. Lakhtakia, Efficiency enhancement of ultrathin CIGS solar cells by optimal bandgap grading, Applied Optics 58 (22) (2019) 6067, https://doi.org/10.1364/ao.58.006067.
[20] M. Ravindiran, C. Praveenkumar, Status review and the future prospects of CZTS based solar cell – a novel approach on the device structure and material modeling for CZTS based photovoltaic device, Renewable & Sustainable Energy Reviews 94 (2018) 317–329, https://doi.org/10.1016/j.rser.2018.06.008.
[21] K. Pal, P. Singh, A. Bhaduri, K.B. Thapa, Current challenges and future prospects for a highly efficient (>20%) kesterite CZTS solar cell: a review, Solar Energy Materials and Solar Cells 196 (2019) 138–156, https://doi.org/10.1016/j.solmat.2019.03.001.
[22] M. Kumar, A. Dubey, N. Adhikari, S. Venkatesan, Q. Qiao, Strategic review of secondary phases, defects and defect-complexes in kesterite CZTS–se solar cells, Energy & Environmental Science 8 (11) (2015) 3134–3159, https://doi.org/10.1039/c5ee02153g.

[23] C. Lee, S. Lee, G.-U. Kim, W. Lee, B.J. Kim, Recent advances, design guidelines, and prospects of all-polymer solar cells, Chemical Reviews 119 (13) (2019) 8028–8086, https://doi.org/10.1021/acs.chemrev.9b00044.
[24] F.C. Krebs, Fabrication and processing of polymer solar cells: a review of printing and coating techniques, Solar Energy Materials and Solar Cells 93 (4) (2009) 394–412, https://doi.org/10.1016/j.solmat.2008.10.004.
[25] P. Roy, N.K. Sinha, S. Tiwari, A. Khare, A review on perovskite solar cells: evolution of architecture, fabrication techniques, commercialization issues and status, Solar Energy 198 (2020) 665–688, https://doi.org/10.1016/j.solener.2020.01.080.
[26] A.K. Jena, A. Kulkarni, T. Miyasaka, Halide perovskite photovoltaics: background, status, and future prospects, Chemical Reviews 119 (5) (2019) 3036–3103, https://doi.org/10.1021/acs.chemrev.8b00539.
[27] T. Matsui, A. Bidiville, K. Maejima, H. Sai, T. Koida, T. Suezaki, M. Matsumoto, K. Saito, I. Yoshida, M. Kondo, High-efficiency amorphous silicon solar cells: impact of deposition rate on metastability, Applied Physics Letters 106 (5) (2015) 053901, https://doi.org/10.1063/1.4907001.
[28] S. Krum, First solar press release, first solar builds the highest efficiency thin film pv cell on record, https://investor.firstsolar.com/news/press-release-details/2014/first-solar-builds-the-highest-efficiency-thin-film-pv-cell-on-record/default.aspx.
[29] M. Nakamura, K. Yamaguchi, Y. Kimoto, Y. Yasaki, T. Kato, H. Sugimoto, Cd-free Cu(In,Ga) (Se,S)2 thin-film solar cell with record efficiency of 23.35%, IEEE Journal of Photovoltaics 9 (6) (2019) 1863–1867, https://doi.org/10.1109/jphotov.2019.2937218.
[30] K. Zhao, Z. Yang, S.F. Liu, Emerging photovoltaic materials and devices, Advanced Functional Materials 29 (47) (2019) 1904014, https://doi.org/10.1002/adfm.201904014.
[31] I. Celik, R.H. Ahangharnejhad, Z. Song, M. Heben, D. Apul, Emerging photovoltaic (PV) materials for a low carbon economy, Energies 13 (16) (2020) 4131, https://doi.org/10.3390/en13164131.
[32] A. Uddin, M. Upama, H. Yi, L. Duan, Encapsulation of organic and perovskite solar cells: a review, Coatings 9 (2) (2019) 65, https://doi.org/10.3390/coatings9020065.
[33] H.S. Dhiman, D. Deb, Wake management based life enhancement of battery energy storage system for hybrid wind farms, Renewable & Sustainable Energy Reviews 130 (2020) 109912, https://doi.org/10.1016/j.rser.2020.109912.
[34] P. Patel, D. Deb, Battery state of charge based algorithm for optimal wind farm power management, in: 2017 6th International Conference on Computer Applications in Electrical Engineering-Recent Advances (CERA), IEEE, 2017.
[35] P. Patel, A. Shandilya, D. Deb, Optimized hybrid wind power generation with forecasting algorithms and battery life considerations, in: 2017 IEEE Power and Energy Conference at Illinois (PECI), IEEE, 2017.
[36] M.A. Eltawil, Z. Zhao, Grid-connected photovoltaic power systems: technical and potential problems—a review, Renewable & Sustainable Energy Reviews 14 (1) (2010) 112–129, https://doi.org/10.1016/j.rser.2009.07.015.
[37] W. Berger, F.-G. Simon, K. Weimann, E.A. Alsema, A novel approach for the recycling of thin film photovoltaic modules, Resources, Conservation and Recycling 54 (10) (2010) 711–718, https://doi.org/10.1016/j.resconrec.2009.12.001.
[38] T.-Y. Wang, J.-C. Hsiao, C.-H. Du, Recycling of materials from silicon base solar cell module, in: 2012 38th IEEE Photovoltaic Specialists Conference, IEEE, 2012.
[39] M.D. Campo, D. Bonnet, R. Gegenwart, J. Beier, Process for recycling cdte/cds thin film solar cell modules, uS Patent 6,572,782 B2, https://patents.google.com/patent/US6572782, 2003.

[40] M.S. Chowdhury, K.S. Rahman, T. Chowdhury, N. Nuthammachot, K. Techato, M. Akhtaruzzaman, S.K. Tiong, K. Sopian, N. Amin, An overview of solar photovoltaic panels' end-of-life material recycling, Energy Strategy Reviews 27 (2020) 100431, https://doi.org/10.1016/j.esr.2019.100431.
[41] E. Klugmann-Radziemska, P. Ostrowski, K. Drabczyk, P. Panek, M. Szkodo, Experimental validation of crystalline silicon solar cells recycling by thermal and chemical methods, Solar Energy Materials and Solar Cells 94 (12) (2010) 2275–2282, https://doi.org/10.1016/j.solmat.2010.07.025.
[42] R. Deng, N.L. Chang, Z. Ouyang, C.M. Chong, A techno-economic review of silicon photovoltaic module recycling, Renewable & Sustainable Energy Reviews 109 (2019) 532–550, https://doi.org/10.1016/j.rser.2019.04.020.

CHAPTER 3

Performance-limiting issues in TFPVs

Thin-film photovoltaic technology has shown a stupendous potential to fulfill the ever-increasing energy requirements of the world. However, there are several bottlenecks in this technology, at both the cell and module levels that hamper its progress. The major issues in thin-film photovoltaics at the cell level are crystal defects, carrier generation, recombination, and mobility degradation. In contrast, the potential induced degradation, dust accumulation, humidity, etc., hamper their performance at the module level. An exhaustive explanation of the aforementioned factors and ways to mitigate their impacts on photovoltaic performance are described in this chapter.

3.1 Introduction

TFPVs have a long history of development, beginning with a power-conversion efficiency of less than 1% to achieving to one of more than 25% [1]. Starting with the advent of single-crystal silicon PV cells, PV technology has indeed come a long way to reach the era of perovskite-based TFPVs [2]. However, issues like stability, durability, toxicity, and recyclability have been the bottlenecks for PV technology's commercialization [3]. Since the material and device-related complexities govern the functionality of TF-PVs, extensive research activities are required in this direction to utilize the full potential of solar energy through photovoltaics. The performance of TFPVs experience various limitations at both the cell and module levels. While phenomena like crystal defects, inferior mobility, carrier recombination, and leakage current are the major performance-limiting factors at the cell level, dust accumulation, the effects of moisture and temperature variation, and potential induced degradation hamper the module-level performance. Some other issues also pose serious implications for TFPV technology's commercialization (Fig. 3.1). This chapter provides a detailed description of the aforementioned performance degrading issues in TFPV cells and modules.

Degradation, Mitigation, and Forecasting Approaches in Thin Film Photovoltaics
https://doi.org/10.1016/B978-0-12-823483-9.00014-0

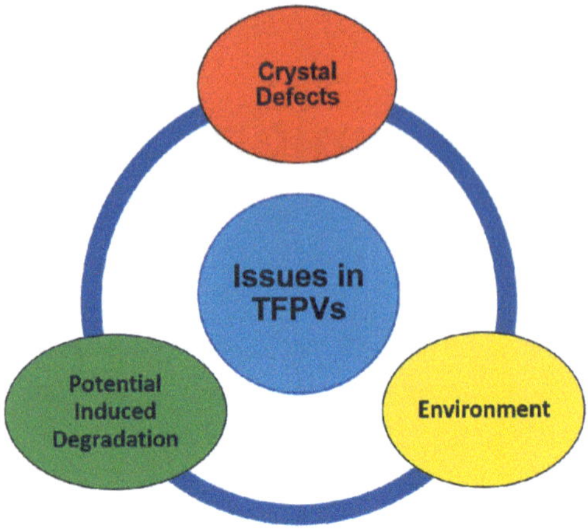

Figure 3.1 Major performance-limiting issues in TFPV technology.

3.2 Crystal defects

The presence of donor/acceptor-type defect states inside the crystal structure severely affects the absorber layer [4]. These defects act as recombination sites for carriers generated post exciton dissociation and thereby limit the efficiency of TFPVs. Moreover, they also produce carrier scattering that restricts the carrier mobility and causing reduced carrier collection efficiency of cells and, hence, conversion efficiency. These facts point towards the importance of understanding the nature of crystal defects to optimally tailor the performance of TFPVs.

In an incessant search for new photovoltaic materials, two naturally important criteria are the magnitude of the bandgap and the visible-range absorption spectrum. The consideration of point-defect processes is equally important. The presence of certain detrimental defects has the potential to cause severe hindrance in generating and transporting photo-generated charge carriers. Despite abundant materials with appropriate band gaps, only countable semiconductor classes have provided more than 10% efficiency. The reason for such a situation is the failure to control the equilibrium concentrations and the presence of efficient nonradiative pathways involving subbandgap electronic states that result from intrinsic defects and extrinsic impurities. Defect tolerance should be a guiding factor in finding novel compounds. Point defects face significantly more difficulty in screening, measuring, and predicting when compared to band gaps.

Thin films of semiconductors suffer from various defect states created inside the crystal structure during the growth process that degrade the overall performance of PV cells. We describe next the commonly observed defect states inside thin films.

3.2.1 Point defects

These zero-dimensional defects are most common in semiconductors, and they act as donors or acceptors depending on their energy level inside the bandgap. Also, they act as trap sites favoring recombination if their energy level is far away from bands. Moreover, they assist in the scattering of carriers during their transport to electrodes. Point defects are created by substituting atoms, removing atoms, or entering an extra atom between lattice sites [5], as shown in Fig. 3.2.

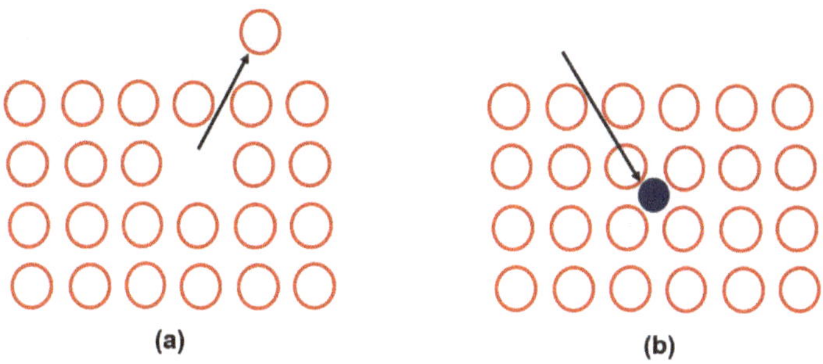

Figure 3.2 Depiction of point-defect formation in crystal: (a) vacancy (b) interstitial.

Many materials are under examination for next-generation solar energy deployments. Such technologies face distinct challenges and opportunities to benefit from point-defect engineering.

3.2.2 Line defects

These one-dimensional defects (also called dislocations) extend throughout the lattice structure along the line of sites. There are two types of line defects, i.e., edge dislocation and screw dislocation (see Fig. 3.3). The edge dislocation happens when an additional half-plane of atoms gets inserted into the middle of a crystal. On the other hand, the screw dislocation is generated when two portions of a crystal slip over each other [6].

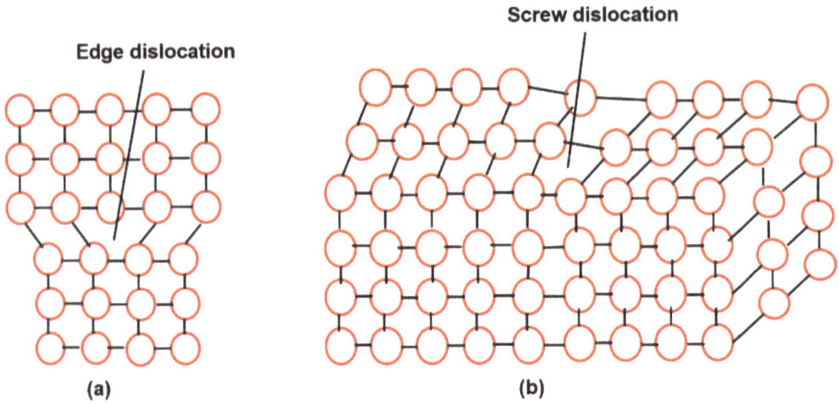

Figure 3.3 Depiction of line-defect formation in crystal: (a) edge (b) screw.

3.2.3 Planar defects

The two-dimensional surface or planar defect originates due to two reasons, i.e., rotation of the lattice without undergoing any change in the long range arrangement of atoms and a face-off between disoriented layers of like or unlike materials. The former category results in twin or stacking faults, while the latter resembles interfaces, surfaces, and grain boundaries inside the crystal (see Fig. 3.4). Twin defects originate due to a rotation of the lattice about an axis perpendicular to the twin plane by 60° or 180°. Stacking faults are generated owing to errors occurring in the stacking of atoms during crystal formation. The planar defects at the surface of a semiconductor layer or between interfaces of different materials arise due to the presence of dangling bonds. In polycrystalline materials, these planar defects give rise to grain boundaries [7].

3.2.4 Volume defects

These three-dimensional defect states can consist of cracks, phase separation, and foreign inclusions. Volume defects are characteristic of the fabrication process, and they generally induce stress throughout the crystal, resulting in distortion of the electronic properties of thin films. However, as a counter strategy, foreign particles are introduced to release the stress and restore the electronic properties (see Fig. 3.5). This process is called 'dispersion hardening' as foreign particles produce hindrances to volume dislocations [8].

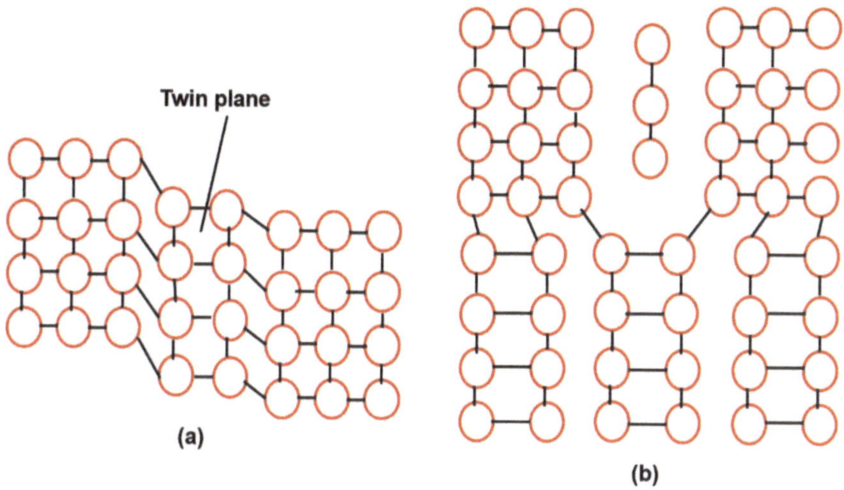

Figure 3.4 Depiction of planar-defect formation in a crystal: (a) twin (b) stacking.

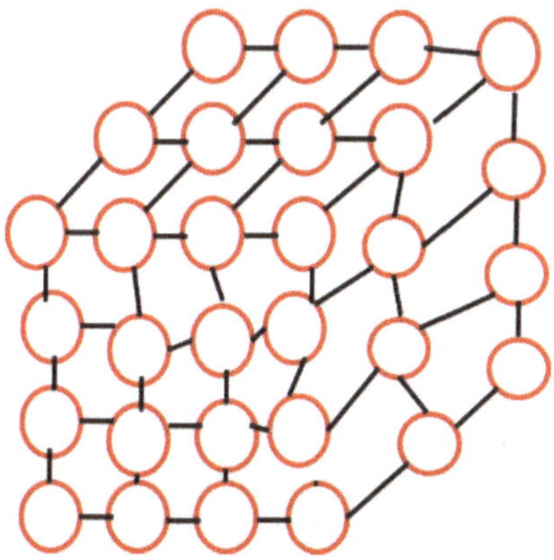

Figure 3.5 Depiction of volume-defect formation in crystal.

3.3 Carrier generation and recombination

In the previous section, we discussed different types of crystal defects that are formed during the growth process of thin films. In this section, we will explain the effect of defect states on carrier generation and recombination processes inside TFPVs [9]. The perturbations occurring inside the absorber layer give rise to the possible electronic states (or defect states) inside the bandgap near or far away from the band edges. For small perturbations (characteristic of dopant atoms), the defect states would probably be located near the band edges (shallow traps). However, with rising perturbations, the defect states will move away from the band edges (deep traps).

Carrier generation is a process in which the population of free carriers (electrons and/or holes) inside the absorber layer increases greatly and abruptly, provided the energy supplied to facilitate the process is greater than the band-gap energy. Generally, in semiconductor crystals the carrier generation process is a result of four basic processes, i.e., thermal generation, impurity-assisted generation, impact ionization, and photo-generation. Thermal generation happens at higher temperatures and in low-band-gap materials. Impurity-assisted generation takes place by interactions with photons or phonons. Impact ionization is caused when carriers moving with high kinetic energy generate additional carriers due to impacts or collisions. Photo-generation is the most important process of carrier generation in TFPVs in which the photons of energy greater than the bandgap of semiconducting absorber layer lead to excess carrier generation.

The process of recombination occurs when an electron is excited from the valence to the conduction band, exists there in a meta-stable state for a very short span of time (nanoseconds or lesser), and falls back to achieve a stable lower energy state inside the valence band. In other words, recombination can be understood as the process of re-establishment of equilibrium that was disturbed during the carrier-generation process. However, in TFPVs it plays a detrimental role that downgrades their performance metrics. Several types of recombination mechanisms occur inside TFPVs, namely, radiative, nonradiative or Shockley–Read–Hall (SRH), and Auger recombination.

The radiative recombination is often observed in direct bandgap semiconductors (e.g., GaAs, CIGS) in which the excited electron present in the conduction band edge directly recombines with a hole present in the valence band edge by releasing a photon of energy equivalent to the bandgap of the semiconductor. The SRH recombination mechanism is characteristic of semiconductors with defect states present in their forbidden gap. In this mechanism, an excited electron in the conduction band will undergo recombination with a hole present in the energy state induced by the defect state present inside the crystal lattice. The Auger recombination process occurs when electrons and holes recombine, however, instead of heat or photon emission, the energy is transferred to a third carrier (electron) in the conduction band, which further gets excited to a higher allowed state in the conduction band and thermalizes back to the conduction band edge.

3.4 Mobility degradation by scattering and grain boundary

The transport of charge carriers generated after the dissociation of excitons is crucial from the perspective of achieving the high performance TFPVs. In this context, the mobility of separated carriers requires important consideration and optimization. Generally, two fundamental mechanisms known as scattering and grain boundary produce a devastating effect on carrier mobility inside the absorber layer, thereby deteriorating the figures of merit in TFPVs [10,11]. The larger the scattering and grain boundaries, the poorer will be the carrier mobility.

In semiconductor crystals, the scattering of carriers and grain boundaries are closely related. Scattering of carriers is aided by phonons and dislocations causing severe degradation of mobility. The scattering of carriers occurs both in-grain (regions inside the crystal where atoms have same orientation) and grain boundary (a transition region between two grains). However, comparatively, the grain boundary has much more profound impact over mobility degradation. The grain boundary produces an energy barrier for carrier movement resulting in the increased probability of recombination, reduced mobility, and therefore the reduced efficiency of TFPVs [12,13].

3.5 Environmental and other factors

Eenvironmental factors, such as dust, moisture, wind speed, bird-poop droppings, and temperature, severely affect the economic benefits obtained from TFPV modules. To some extent, these factors are responsible for restricting the huge investment prospects in developed and developing countries. The aforementioned environmental factors degrade the performance of TFPVs, both at the cell and module levels. Therefore, some counter strategies are required to mitigate the environmental effects on TFPV technology.

Especially in those geographical locations where these environmental factors are mainly responsible for the unsustainable functioning of TFPV panels, the mitigation strategies should be compulsorily adopted. Although most of these factors are simultaneously acting, it is important to segregate and investigate the extent of degradation caused by each factor. In this section, we will discuss all the major environmental factors responsible for the performance degradation of TFPV panels.

3.5.1 Effect of dust accumulation

Dust is the collection of small solid particles typically of a diameter of 500 μm or less. These also can include bacteria, fungi, and microfibers of fabrics. Several environmental and climatic conditions assist the natural phenomenon of dust accumulation. Moreover, gasoline-powered vehicles, construction activities, pollution, volcanic eruptions, and wind also contribute to scattering dust into the atmosphere [14]. The amount of dust accumulation over surfaces is governed by the dust properties (size, shape, chemical properties, weight, etc.), climatic conditions (site-specific weather conditions like wind speed and humidity), and surface features (module tilt angle and surface finish). Several studies have attempted to investigate the effect of dust accumulation on TFPV cells and modules. A broad range of performance reduction has been reported, depending on the amount of dust accumulated. The primary reason causing the deterioration of TFPV systems' deterioration upon dust accumulation is the reduced extent of sunlight penetration through the TFPV cell matrix (loss of transmittance) [15]. A common strategy generally adopted to reduce dust accumulation is to install the TFPV panels in a tilted arrangement (finite inclination). We can easily verify the efficacy of a tilt angle to reduce the effect of dust accumulation through results obtained for TFPV installations over time at two significantly different values of tilt angle.

The panels installed at a larger tilt angle result in a relatively lesser extent of performance degradation. As far as the size of dust particles is concerned, TFPV installation's efficiency is more affected by finer particles' accumulation than coarser particles. An important observation about the effect of dust accumulation on panel installations is that "dust promotes dust," which causes an exponential decrease in TFPV performance. In this context, wind patterns could be helpful in the reduction of dust accumulation. However, high humidity further promotes dust coagulation, reducing the role of wind speed in dust removal from the panel surface. The influence of wind speed is both positive and negative towards dust accumulation. Analysis shows that dust accumulation reduces at low wind speeds, but increases at high wind speeds since the greater spreading of dust particles causes more layer deposition.

Some studies have also pointed towards the greater importance of wind direction over wind speed. The phenomenon of dust accumulation is usually reduced by cleaning procedures like forced airflows, natural cleaning with rainwater or wind, forced-water cleaning, mechanical cleaning, manual cleaning, electrodynamic display, superhydrophobic aircraft, and self-ultrasonic cleaning. [16].

3.5.2 Effect of humidity

Similar to dust accumulation, humidity is an essential environmental factor responsible for variations in TFPV performance. The moisture present in the air surrounding the installation affects the output in two ways. First, the minute water droplets in the atmosphere profoundly impact the irradiance level received by panels because sunlight is reflected, refracted, and diffracted depending on the size of the water droplet. This effect causes small variations in 'V_{OC}' but large variations in 'J_{SC}' [17]. Secondly, the penetration of moisture inside the TFPV cells causes delamination of the encapsulation layer. The varying sizes of water droplets and their distribution in the atmosphere are responsible for the nonlinear variation of irradiance level. The high density of small water-vapor particles causes large scattering and diffraction of sunrays resulting in a sharp drop in the irradiance level.

Also, corrosive moisture ingression inside TFPV modules causes the failure of cell interconnections or even the cracking of cells, and further expansion of deterioration happens under hot and humid conditions [18]. Therefore, it is essential to improvise novel designs of TFPVs to achieve sustainable power output under high relative-humidity conditions. The rising humidity levels surrounding a TFPV installation lead to sticking of dust to the module's top glass surface, which severely degrades the output. Therefore, frequent rigorous cleaning is required to obtain a sustainable outcome from the TFPV panels. The problem of dust adhesion on TFPV installation can be ameliorated by applying a hydrophilic (water-attracting) antisoiling coating over the top glass surface. The antisoiling coating [19] improves the transmittance of the top surface of the TFPV module directly facing the sun, provided its surface energy is sufficiently high and at a contact angle of less than 40°.

Alternatively, the hydrophobic anti-soiling coating can also be employed to make water droplets accumulate and then slide off the inclined panel's surface. Generally, the solar industry prefers hydrophobic coatings in drier and polluted locations subject to a broad range of dust particles and organic contaminants. The hydrophilic coating is formed by to a uniform film of water covering the panels' surface, and the deposited dust particles get washed away with excess moisture (such as heavy rainfall). However, both these innovative approaches have witnessed limited success.

3.5.3 Effect of wind

The wind crossing the installation is important from the perspective of lowering the module temperature and thereby maintaining its performance metrics [20].

Generally, it has been observed that module temperature (relative to ambient temperature) can be reduced to half when wind speed is increased to more than 10 m/s. Hence, wind speed plays a crucial role in maintaining the performance of modules, although the direction of wind has been witnessed to have an insignificant impact on module temperature. Additionally, the surface structure of the TFPV module has a significant role to play in maintaining its temperature. At reasonably high wind speeds, the rise in temperature of modules is controlled through convective cooling. In this context, the structured glass modules prove to be much more efficient than flat ones. According to a finding, the temperature of grooved glass module was observed to be 3.5°C lower than a flat glass cover module, although with a sufficiently high wind speed.

3.5.4 Effect of temperature

The TFPV modules, irrespective of their conversion efficiency, utilize only a small part of the solar spectrum to generate electricity, while most of the unutilized radiation merely raises the temperature of the module, causing efficiency reduction [21]. Therefore, it becomes necessary to enquire about the temperature coefficient and efficiency of a module before purchasing it, and many a times these performance metrics are different from the values claimed by the manufacturers. This highlights the important of the pretesting of modules under simulated environmental condition of variable temperatures. The rising temperature inside a TFPV module tends to lower the band gap of cells, causing an increment of the minority carrier lifetime and the absorption of lower-energy photons. The effect is reflected in terms of a slight increase in 'J_{SC}' but a significant reduction in 'V_{OC}', thus producing an overall decrease in the fill factor. To assess the performance stability of TFPV modules under rising temperatures, the temperature coefficient is measured in the following way. The TFPV module is heated to a predetermined temperature (typically 80°C) and I-V characteristics are measured inside the solar simulator. The same measurement was repeated repeatedly at several lower values of temperature until the module is cooled down to the ambient temperature (25°C). Also, the trend of efficiency dependence on temperature was observed to be a function of the cell and encapsulation material. Specifically, the absorptive and dissipation properties of the encapsulating material prove to be helpful in mitigating the heating effect produced by rising temperatures.

3.5.5 Effect of module orientation

The inclination of TFPV module determines the amount of sunlight received by it. The maximum conversion efficiency has been observed to be greatly affected by the module's tilt angle [22]. Thus, it is important to obtain an optimal tilt angle during the installation of TFPV modules. The initial finding indicates that an optimal tilt angle can be easily estimated using the relation: "latitude ±15°" [23]. Later, a more accurate relation was formulated: "latitude ±8°". In modern times, the TFPV solar-tracking technique has evolved quickly that helps in avoiding the overheating of modules, although the overall cost of the whole system increases, due to an increased initial and running costs, leading to a increase in the energy payback period. Therefore, there is no doubt that inclusion of a tracking system enhances the overall efficiency of the system but at thge expense of reduced economic feasibility. Of the two types of tracking systems, the single-axis tracking system is a more convenient and economical alternative to a dual-axis tracking system.

3.5.6 Effect of shading

The performance metrics of TFPV installation is severely affected due to complete or partial shading. The effect of shading depends on module position, the shading scenario, and the position of module [24]. The most common sources of complete/partial shading are trees, dust accumulation, buildings, bird droppings, and self-shading caused by other existing TFPV modules. All these shading effects can be minimized through proper preplanning achieved by careful analysis of the installation site. This can be achieved by optimizing the solar-elevation angle, inclination angle, and minimal distance. The effect of shading is also evident in terms of fluctuations in the power-voltage characteristics and the development of advanced models are required to track the performance of shaded TFPV panels. Some popular models for analysis of the shading effect are the bypass diode, Lambert *W* function, two stages successive estimation, and Bishop models.

3.6 Leakage current and potential induced degradation

The electrical power output of a single TFPV cell is generally insufficient to drive any significant load. Thus, many cells connected in series (modules) help achieve voltage enhancement at a constant current value (low ohmic losses). These cells are also electrically isolated or insulated from current-carrying parts or grounded frames. However, leakage currents flow between active cell-matrixes and the ground, causing several TFPV degradations such as electrochemical corrosion of cells, interconnects and frame, and potential induced degradation (PID) of cells [25]. The mentioned TFPV degrades more quickly with an increased leakage current.

The major contributor to leakage current in PVs is the potential difference between the active cell-matrix and the module. PID refers to the migration of positive ions (e.g., Na^+ ions) from the glass substrate to the cell via encapsulation and antireflection coating, thereby causing surface polarization and shunting [26]. The effect of PID is generally evident after months or years of installation when these ions cross various bulks and interfaces inside the cell, which result in the degradation of the overall power output of the module. The probable paths of leakage current inside a TFPV module are surface and/ or bulk of front glass and encapsulant (Path-1), the bulk of front glass and bulk of encapsulant (Path-2), the interface of front glass/encapsulant, and the bulk of encapsulant (Path-3), the bulk of encapsulant (Path-4), interface of encapsulant/backsheet and the bulk of encapsulant (Path-5), and the surface of backsheet and bulk of backsheet and encapsulant (Path-6) [27].

Among these, Path-1 is the most deteriorating factor due to an enhanced conductivity of front glass in the presence of rain or humidity. Path-6 often is ignored due to the high electrical resistivity of the polymer backsheet. The phenomenon of PID has attracted a considerable amount of research interest in recent times due to to the gradual but disastrous degradation of TFPV performance, subsequently resulting in unreliable longevity of modules. Initially, PID was observed to occur inside the crystalline silicon PV modules, but recently, its presence has also been confirmed in TFPV modules [28]. Although extensive research is being carried out on PID's effect, understanding of it remains insufficient. It has been quite a challenge to understand the complexity of PID as because it involves several variables, such as the properties of cell material, encapsulation material, anti-reflection coating, module construction, environmental conditions like humidity, temperature, and dust accumulation.

The root cause of PID varies with module technology and/or operating conditions, and both macroscopic and microscopic investigations are often useful for its assessment. Past studies on PID have suggested that the shunt resistance of TFPV modules undergoes drastic reduction and an increased dark current due to enhanced recombination inside the cells. The movement of sodium ions (Na^+) is the precise reason behind the PID phenomenon. The ions penetrate the encapsulation layer and anti-reflection coating to enter the cell structure and subsequently produce defect states at p–n junctions, resulting in an increased recombination center population that yields a degraded power conversion efficiency (Fig. 3.6(a)).

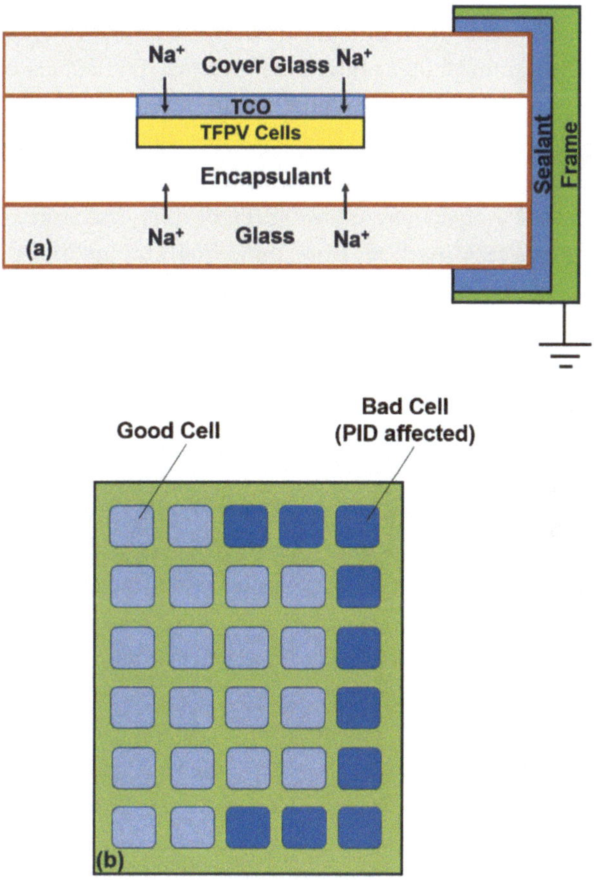

Figure 3.6 Schematic depiction of sodium-ion movement inside TFPV module (b) PID effect damaging the TFPV cells in a module.

The effect of PID is more significant at a low level of irradiance because carrier losses due to recombination become dominant (reduced shunt resistance), leading to a reduced extent of photocurrent generation. The impact of PID is due to the movement of Na^+ ions in the negative potential to the ground. Experimentally, we can visualize PID using Scanning Electron Microscopy (SEM), Electron Beam Induced Current measurement (EBIC), Time-of-Flight Secondary Ion Mass Spectroscopy (ToF-SIMS), or Dark Lock-in Thermography (DLIT). The movement of Na+ ions originates from the soda-lime glass sheets, which contain 10–14% Na_2O and have a resistivity of approximately 1011 Ω-cm at room temperature. The effect creates stacking faults inside the absorber layer's bandgap, thus producing Shockley–Read–Hall (SRH) recombination sites. Also, it is common to observe that cells near the frame are more affected by PID compared to those present in the module's center (Fig. 3.6(b)). The effects of PID progression happening inside TFPV modules are identified using several test methods. Since PID happens both at cell and module levels, it becomes necessary to investigate PID progression at both levels [29]. PID at the cell level is generally characterized using the corona-discharge method and module-like layer-stack method. Both these methods are useful in probing the root cause of PID in TFPV cells.

In the corona discharge test, positive ions are transferred through the tip of a thin wire (connected to a high potential) to the test sample's front end. An electric field is induced across the test sample. In this case, the accumulation of charges cannot be considered analogous to a pile of Na^+ ions, and also, the ion discharge sometimes damages the test sample. Again, this technique ignores the variation in electric-field distribution induced by the encapsulant and cover glass. The module-like layer-stack method was developed to counter the drawbacks of the corona-discharge method because it acknowledges the impact of packaging materials during testing of PID progression inside TFPV cells. The progression of PID in the TFPV cells is shown with greater details in Fig. 3.7.

In this technique for PID testing, the test TFPV sample is placed on a temperature-controlled uniformly heated aluminum platform. To imitate a TFPV module scenario, the sample TFPV cell's top surface is covered with encapsulant and a glass sheet (Fig. 3.8). Subsequently, a suitable voltage is applied across the test sample. This experimental setup prevents the contamination of the test sample and can be isolated from encapsulation and glass sheet after measurement of PID.

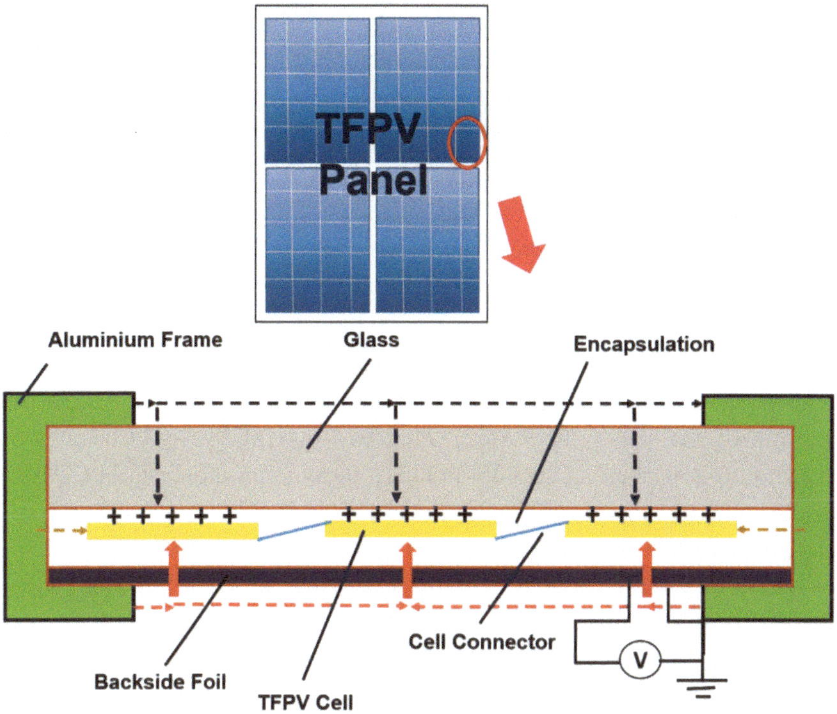

Figure 3.7 Leakage current path inside the TFPV cell array.

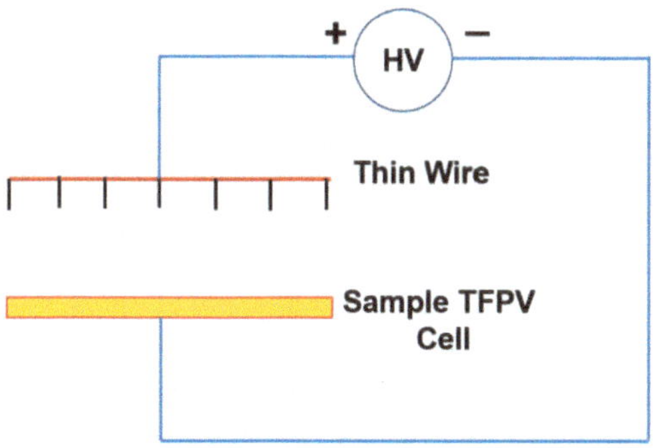

Figure 3.8 Experimental setup for the corona-discharge method of PID testing for TFPV cell.

The results obtained with this technique are more informative and reliable because the in-situ measurement of shunt resistance is possible, thus yielding an accurate assessment of PID progression in TFPV cells. PID testing at the module level is done using two methods. The first method is known as the climate chamber technique.

The TFPV test module is placed under controlled high temperature and humidity conditions, maintained inside a chamber, with high electrical potential applied across it. High precision is required while controlling the temperature and humidity levels inside the chamber to achieve the test results' repeatability since both these factors have a decisive impact on PID progression. In particular, the humidity fluctuation has to produce additional stress inside TFPV modules, which is speculated to be due to condensation of moisture content over the front glass when the module temperature lags too much behind the chamber temperature (due to severe ramping of chamber temperature). Hence, a quantitative optimization of temperature ramping is generally required to maintain the accuracy of test results so that the ramping time can be minimized by avoiding condensation at the same time.

In the second method, PID testing is done by placing a conductive layer (copper/aluminum foil or carbon paste) over the front glass surface of the TFPV module under test (Fig. 3.9). This method is also called the "aluminum foil PID test" since aluminum foil is generally used to cover the top of the front glass surface, and this method works as an alternative to chamber based PID testing. The reason for placing a conductive foil on the front glass surface is to mimic the PID effect produced under highly humid conditions since the foil also provides a conductive path on the glass surface. The incorporation of aluminum foil reduces the humidity effort, although temperature control is still required.

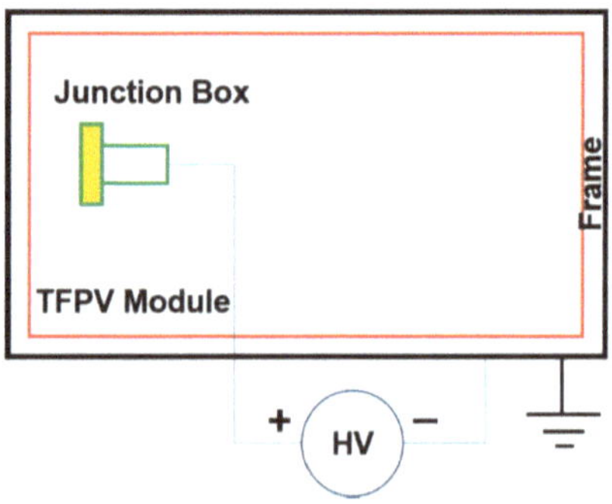

Figure 3.9 Experimental setup for climate-chamber method for PID testing TFPV modules.

For perfect transfer of stress from aluminum foil to glass surface, proper contact between the two is a must and is generally achieved by placing a small rubber mat over the aluminum foil. This PID progression testing method offers advantages like a smaller requirement to control humidity and considerable more performance degradation under similar conditions of applied stress. However, with the chamber PID test performed under a natural environment-like conditions (high humidity and temperature conditions), it proves to be more field relevant.

Bibliography

[1] M.A. Green, The path to 25% silicon solar cell efficiency: history of silicon cell evolution, Progress in Photovoltaics: Research and Applications 17 (3) (2009) 183–189, https://doi.org/10.1002/pip.892.

[2] T.D. Lee, A.U. Ebong, A review of thin film solar cell technologies and challenges, Renewable & Sustainable Energy Reviews 70 (2017) 1286–1297, https://doi.org/10.1016/j.rser.2016.12.028.

[3] A. Urbina, The balance between efficiency, stability and environmental impacts in perovskite solar cells: a review, Journal of Physics: Energy 2 (2) (2020) 022001, https://doi.org/10.1088/2515-7655/ab5eee.

[4] S. Joonwichien, S. Matsushima, N. Usami, Effects of crystal defects and their interactions with impurities on electrical properties of multicrystalline si, Journal of Applied Physics 113 (13) (2013) 133503, https://doi.org/10.1063/1.4798600.

[5] J.S. Park, S. Kim, Z. Xie, A. Walsh, Point defect engineering in thin-film solar cells, Nature Reviews Materials 3 (7) (2018) 194–210, https://doi.org/10.1038/s41578-018-0026-7.

[6] G. Rajasekaran, P. Narayanan, A. Parashar, Effect of point and line defects on mechanical and thermal properties of graphene: a review, Critical Reviews in Solid State and Materials Sciences 41 (1) (2015) 47–71, https://doi.org/10.1080/10408436.2015.1068160.

[7] M.D. Rossell, P. Agrawal, M. Campanini, D. Passerone, R. Erni, Atomic structure and electronic properties of planar defects in $SrFeO_{3-\delta}$ thin films, Physical Review Materials 4 (7) (2020), https://doi.org/10.1103/physrevmaterials.4.075001.

[8] L.E. Murr, Volume defects: 3d imperfections in crystals, in: Handbook of Materials Structures, Properties, Processing and Performance, Springer International Publishing, 2015, pp. 313–324.

[9] S. Shoaee, M. Stolterfoht, D. Neher, The role of mobility on charge generation, recombination, and extraction in polymer-based solar cells, Advanced Energy Materials 8 (28) (2018) 1703355, https://doi.org/10.1002/aenm.201703355.

[10] B. Olyaeefar, S. Ahmadi-Kandjani, A. Asgari, Classical modelling of grain size and boundary effects in polycrystalline perovskite solar cells, Solar Energy Materials and Solar Cells 180 (2018) 76–82, https://doi.org/10.1016/j.solmat.2018.02.026.

[11] H. Zhang, G. Lee, C. Gong, L. Colombo, K. Cho, Grain boundary effect on electrical transport properties of graphene, The Journal of Physical Chemistry C 118 (5) (2014) 2338–2343, https://doi.org/10.1021/jp411464w.

[12] Z. Chu, M. Yang, P. Schulz, D. Wu, X. Ma, E. Seifert, L. Sun, X. Li, K. Zhu, K. Lai, Impact of grain boundaries on efficiency and stability of organic-inorganic trihalide perovskites, Nature Communications 8 (1) (2017), https://doi.org/10.1038/s41467-017-02331-4.

[13] I. Vladimirov, M. Kühn, T. Geßner, F. May, R.T. Weitz, Energy barriers at grain boundaries dominate charge carrier transport in an electron-conductive organic semiconductor, Scientific Reports 8 (1) (2018), https://doi.org/10.1038/s41598-018-33308-y.
[14] T.M. Alnasser, A.M. Mahdy, K.I. Abass, M.T. Chaichan, H.A. Kazem, Impact of dust ingredient on photovoltaic performance: an experimental study, Solar Energy 195 (2020) 651–659, https://doi.org/10.1016/j.solener.2019.12.008.
[15] S.A. Said, G. Hassan, H.M. Walwil, N. Al-Aqeeli, The effect of environmental factors and dust accumulation on photovoltaic modules and dust-accumulation mitigation strategies, Renewable & Sustainable Energy Reviews 82 (2018) 743–760, https://doi.org/10.1016/j.rser.2017.09.042.
[16] H.A. Kazem, M.T. Chaichan, A.H. Al-Waeli, K. Sopian, A review of dust accumulation and cleaning methods for solar photovoltaic systems, Journal of Cleaner Production 276 (2020) 123187, https://doi.org/10.1016/j.jclepro.2020.123187.
[17] N.C. Park, W.W. Oh, D.H. Kim, Effect of temperature and humidity on the degradation rate of multicrystalline silicon photovoltaic module, International Journal of Photoenergy 2013 (2013) 1–9, https://doi.org/10.1155/2013/925280.
[18] M.D. Kempe, A.A. Dameron, M.O. Reese, Evaluation of moisture ingress from the perimeter of photovoltaic modules, Progress in Photovoltaics: Research and Applications 22 (11) (2013) 1159–1171, https://doi.org/10.1002/pip.2374.
[19] A.J. Labelle, N. Coish, M. Sinclair, J. Slonetsky, S. Myrskog, J.P. Morgan, Anti-soiling coatings for sun simba concentrated photovoltaic (CPV) modules, Author(s), 2016, https://doi.org/10.1063/1.4962080.
[20] C. Schwingshackl, M. Petitta, J. Wagner, G. Belluardo, D. Moser, M. Castelli, M. Zebisch, A. Tetzlaff, Wind effect on PV module temperature: analysis of different techniques for an accurate estimation, Energy Procedia 40 (2013) 77–86, https://doi.org/10.1016/j.egypro.2013.08.010.
[21] S. Dubey, J.N. Sarvaiya, B. Seshadri, Temperature dependent photovoltaic (PV) efficiency and its effect on PV production in the world – a review, Energy Procedia 33 (2013) 311–321, https://doi.org/10.1016/j.egypro.2013.05.072.
[22] A. Lahnaoui, P. Stenzel, J. Linssen, Tilt angle and orientation impact on the techno-economic performance of photovoltaic battery systems, Energy Procedia 105 (2017) 4312–4320, https://doi.org/10.1016/j.egypro.2017.03.903.
[23] H. Darhmaoui, D. Lahjouji, Latitude based model for tilt angle optimization for solar collectors in the Mediterranean region, Energy Procedia 42 (2013) 426–435, https://doi.org/10.1016/j.egypro.2013.11.043.
[24] J.P.N. Torres, S.K. Nashih, C.A.F. Fernandes, J.C. Leite, The effect of shading on photovoltaic solar panels, Energy Systems 9 (1) (2016) 195–208, https://doi.org/10.1007/s12667-016-0225-5.
[25] S. Voswinckel, P. Manz, C. Schmidt, V. Wesselak, Investigation of leakage currents depending on the mounting situation in accordance to amorphous silicon modules, Energy Procedia 57 (2014) 56–64, https://doi.org/10.1016/j.egypro.2014.10.008.
[26] G. von Gastrow, E. Martinez-Loran, J. Scharf, J. Clenney, R. Meier, P. Bandaru, M.I. Bertoni, D.P. Fenning, Quantification of sodium-ion migration in silicon nitride by flatband-potential monitoring at device-operating temperatures, Physica Status Solidi A 217 (16) (2020) 2000212, https://doi.org/10.1002/pssa.202000212.
[27] N.G. Dhere, N.S. Shiradkar, E. Schneller, Device for detailed analysis of leakage current paths in photovoltaic modules under high voltage bias, Applied Physics Letters 104 (11) (2014) 112103, https://doi.org/10.1063/1.4869028.
[28] W. Luo, Y.S. Khoo, P. Hacke, V. Naumann, D. Lausch, S.P. Harvey, J.P. Singh, J. Chai, Y. Wang, A.G. Aberle, S. Ramakrishna, Potential-induced degradation in photovoltaic modules: a critical review, Energy & Environmental Science 10 (1) (2017) 43–68, https://doi.org/10.1039/c6ee02271e.

[29] S. Mishra, K. Bhargava, D. Deb, Numerical simulation of potential induced degradation (PID) in different thin-film solar cells using SCAPS-1d, Solar Energy 188 (2019) 353–360, https://doi.org/10.1016/j.solener.2019.05.077.

CHAPTER 4

Yield increase through soiling prevention

Solar energy is available in abundance, and it is possible to extract it by using solar cells and through regular solar-panel maintenance. However, solar cells' relatively low efficiency is further affected by factors such as the orientation of the panel, shade, wind speed, ambient temperature, precipitation, and dust deposition. Significantly, the maintenance of solar panels is challenging on account of soiling. Currently, the usual method for removing debris from PV panels is manual cleaning with water-based cleaning solutions. This chapter reviews currently employed methodologies, including mitigation techniques, to help users choose the most appropriate method per the location and the available resources for dust mitigation for improved yield.

4.1 Introduction

Solar energy is an efficient renewable resource due to its wide availability and ease of generation, but a solar panel's yield invariably reduces over time due to soiling. Accumulation of dust on photovoltaic modules' surfaces decreases the incoming irradiance to the solar cell, obscures solar flux, and causes a power loss. Studies in dry areas show that these losses could reach up to 15% of the total capacity generated per day. In such cases, the typical solution is to clean the panels with water. In large-scale PV plants, water-based cleaning is expensive, especially when there is a limited water supply. As a result of being exposed to external conditions, the PV system encounters dust formation, affecting its system performance [1]. The term dust generally refers to any atmospheric particle with a diameter of less than 500 μm [2].

Given that we expect solar PV installations to provide 25% of the world's total electrical power needs, the PV installations are more likely to be in nonagricultural or semiarid regions and deserts, which together cover a third of the world's landmass. Projections indicate that, even the installation of PV plants in 4% of the net viable arid regions, the total global-energy

Degradation, Mitigation, and Forecasting Approaches in Thin Film Photovoltaics
https://doi.org/10.1016/B978-0-12-823483-9.00015-2

requirement is attainable with close to zero CO_2 emissions, in contrast to the alarming levels of greenhouse gases that increase global warming [3].

Almost half of the solar irradiance is scattered and absorbed in the atmosphere before striking the Earth's surface. The annual amount of solar radiation is concentrated in arid regions near 25° N and S 25° S of the Equator and falls off sharply close to the Equator, or the poles [4]. In these regions, the cloud cover considerably contributes to the variation of insolation. Available data about the mean annual global irradiation, measured in kWh's energy units (1 kWh = 3.6 × 106 J), identifies that the highest incidence occurs in the Sahara Desert at 2685 kWh/m^2/year, followed by The Great Sandy Desert and the Thar Desert. Fossil fuels contribute about 90% of the currently estimated energy requirement, that is, 474 × 1018 J. Assuming a future world with about 10 billion people, each using an average of 10 kW, a total power of 100 TW would be needed. The expected energy needs are achievable by installing PV power panels with around 14% generation efficiency in a landmass of approximately 3.5 × 105 km^2. A small fraction of the total landmass in the seven significant deserts vegetation cannot grow [5].

Soiling is one of the biggest challenges for PV in desert environments. In arid regions worldwide, we typically see soiling rates on the order of 0.1 to 1% a day. We take a mid-range value of .5%, which is what the daily average in Doha, for instance. After one month without cleaning, a PV system's power falls by approximately 15% just because of the dust. Putting that into perspective, over the last 20 years, the efficiency of silicon PV modules has increased from roughly 15% up to 20%. So, if you think about it, it only takes about two months of soiling to wipe out two decades of improvements in PV module technology. Suspended dust (in the atmosphere) in the direct optical path of an incident irradiance of solar panels and also the dust deposited on the panels both obscure the impending solar radiation, particularly in dusty, arid, or semiarid regions, or near highways, significantly impacting solar panels' power-capture capacity. Goossens et al. report that large amounts of dust are generally deposited on panels due to powerful westerly winds in the afternoons and southwesterly winds in the evenings [6]. Remote measurements of suspended atmospheric particulate matter made in many different regions indicate that higher light-obstruction levels originate from particles of diameter 0.3–0.6 mm. These sizes are similar to the visible-light wavelength. The rate of dust deposition widely varies. For instance, some desert regions in the Middle East have a 0.36 g/m^2 rate of dust deposition per day, whereas 0.5 g/m^2 a day in the Negev Desert and 0.17 g/m^2 a day in the Mojave Desert have been reported [7].

Desert storms followed by light rains are common phenomena that result in the formation of a thin layer of mud on the exposed PV panel surface. After the storm, efficient cleaning of the dust layer helps remove this barrier to the incident irradiance. Many such PV panels are cleaned with water, a costly, labor-intensive process, especially when clear water is scarce. Most PV panels are fabricated from tempered borosilicate glass, which is easy to clean with water. In recent advances, textured glass helps trap light and lower reflection losses, but such textured surfaces also attract and deposit microscopic particles, so therefore they need efficient cleaning.

How do we measure soiling loss in the field? The most straightforward way is to keep one PV module clean, allow the other to become soiled, and measure the difference in their power outputs. In the middle of the day, when the sunrays are incident in a more-or-less typical way to the module, we have a certain amount of dust and a wide beam of irradiation. But in the morning and afternoon, the same amount of dust blocks a much narrower beam of irradiation. So, the measured soiling loss increases as we move away from noon.

As of now, there are no magic bullets that stop dust from accumulating on PV modules. So-called antisoiling coatings are available on the market, but at best, they slightly reduce the soiling rate. In other words, they increase the time between cleanings. Also, their performance depends significantly on the local climate. For example, whether or not there's condensation at night. So the key message is to test an anti-soiling coating in local conditions before committing to it on a large scale for a particular project. Electrodynamic dust shields are also under development that are intended to repel dust using electric fields. They're still in the research stage and are far from commercial deployment.

For the foreseeable future, we have to clean our PV systems somehow. Machines are now available on the market, running along the PV arrays to eliminate manual labor from the cleaning. They're still rare in commercial PV plants, but they are beginning to be adopted in some places. The critical question for cleaning machines is whether they cause abrasion of the module coatings, and testing standards are being developed to help evaluate the abrasion. Right now, by far, the dominant cleaning method in desert regions is still manual cleaning. It's reliable; it's useful. However, there are high labor costs. That raises the critical question: What is the optimal schedule for manual cleaning of PV systems? How often should we clean? There's a trade-off between the cost of cleaning and the benefit of enhanced energy generation. One could clean every day, but that would cost more than the value of the electricity yield. The exact optimum or sweet spot depends on each project's unique factors, such as labor cost, the electricity price, the efficiency of the modules, and the soiling rate.

Factors governing dust accumulation on PV panels include: (i) wind velocity, (ii) ambient temperature, (iii) humidity, (iv) locational characteristics like vehicular traffic and air pollution, (v) the orientation and tilt angle of the PV panels, (vi) Coating characteristics and texture, and (vii) the type of dust (size, shape, and weight). In this book, we deal with these factors in varying degrees.

4.2 Measuring output of PV cells, modules; and systems

The energy produced by a panel depends greatly on the weather conditions like insolation, temperature, and wind speed. Therefore, it's critical to monitor these parameters to interpret the modules' performance in the field accurately. Weather stations include pyranometers that measure the global irradiance and pyrheliometers for the direct normal irradiance. Also, anemometers measure the wind speed, and hygrometers monitor the humidity. There are also tools for indoor and outdoor testing of cells and modules. These include a solar simulator using a xenon arc lamp as the light source. However, recently, laser sources and light-emitting diodes have also been tested to generate spectra that match the solar spectrum.

Portable IV testers provide IV curves to determine all parameters that characterize the cell and module, such as the series resistance and the shunt resistance. The low-voltage part of the curves is quite sensitive to the shunt resistance, while the high-voltage part is susceptible to series resistance. The module performance reported in the datasheet corresponds to the standard conditions, 25°C and 1000 watts per square m irradiance. Sun simulators can offer the possibility to vary the incident power. The open-circuit voltage increases logarithmically as the incident power rises, while the short-circuit current increases linearly. However, we observe some deviation from the linearity at low irradiance. The module efficiency is usually lower than the cell efficiency. Many factors contribute to the additional power losses at the module level—the most important of them being further losses in the interconnections.

How do we measure PV performance in the field, and what instruments do we need? Two standard calculations we can make are the specific yield (daily energy produced divided by nominal power) and performance ratio. The specific yield is ideal for comparing how much energy we get from one kind of module versus another at a given location. When we buy modules, we're paying for their power rating or watts-peak. Still, when we sell their electricity, we're selling kw-hr, so it's critical to know how many kw-hrs we get per kilowatt peak, and that's what specific yield is.

There's no fixed specific yield number for a module. It's not printed on the datasheet but depends on local conditions like sunlight and temperatures. Comparing two modules, the first might have a higher specific yield in a cold climate, while the second might be better in a hot climate. It's essential to test them side-by-side at the actual intended location to determine the specific yield numbers for two different modules. Overall, we see more variation between modules of a given technology type than between the averages of other technologies. So the takeaway from this data is that no PV technology class on the market is inherently unsuited for deserts. And, it's hard to predict how a module will perform just from its datasheet. If one wants to know with confidence how many kw-hr are obtainable from each kilowatt peak of modules, it's essential to field test them under local conditions.

We often aim to monitor a PV system's performance over time to see if it degrades or develops any faults. A useful metric for this is the Performance Ratio. It tries to correct all the "normal" factors that influence module power, so the only remaining variation is to examine "abnormal" aspects that interest us. When we say "normal" elements, we're talking about mainly irradiation and module temperature. Remember that the power rating, or watt-peaks of a module, is how much power it produces at standard test conditions of a 1000 W/m^2 at 25°C. So, nominally, with the module temperature of 25 divided by irradiation and the power rating, you should get the number one. In reality, the Performance Ratio measures how far the real PV system has deviated from that nominal value of one. So the Performance Ratio is "immune" to changes in temperature and irradiation. The main remaining contributing factor is spectrum variation.

The other thing one notices is that, even when the system is clean, the Performance Ratio is close to but not precisely at the nominal value of one. This is typical of real PV systems in the field and doesn't indicate a problem by itself. What we're interested in is how that Performance Ratio changes the over time. If we see a significant drop in the clean Performance Ratio, this would indicate a fault somewhere. Since the modules are not going to be at 25°C in the desert, we need to use the module's power-temperature coefficient as listed on its datasheet. We use that to adjust the power to the level it would have been at if we could somehow magically keep the module at 25°C. If we didn't do this, we would be overlaying temperature effects on top of the abnormal results, obscuring the degradation.

As winter turns into spring and the temperature increases, the Performance Ratio drops. You can also see that this introduces more spikiness due to daily differences in module temperature. We need to know the module temperature in the field to adjust the power from the actual module temperature to 25°C. In the Performance Ratio equation, we need to know the plane of the array of irradiation. So, unlike the Specific Yield method we saw earlier, which only required the module power, we also need to measure the module temperature and irradiation for the Performance Ratio. Because we'll be making these measurements over decades, it's worth investing in a high-quality data-acquisition system upfront.

Existing solutions for soiling mitigation include manual and semi-automated cleaning methods that require water in excess amounts and are labor-intensive, thereby further decreasing their cost-effectiveness and eco-friendliness. Some autonomous solutions wherein an electrodynamic screen (EDS), placed atop a PV panel, removes dust with the help of an electric field, generated from a high voltage supply to the EDS electrodes, are discussed further [8]. These screens may not always be efficient due to their heavy weight and uneconomical required services, thereby reducing commercial feasibility. In general, cleaning the PV panel surface has only been emphasized recently because rainfall in PV installation regions were previously considered adequate to clean the PV panel surface. However, recent studies prove that soiling can severely affect PV panels' energy yield even in areas receiving sufficient rainfall.

4.3 Methods of cleaning solutions

From a review of the literature like [9] and [10], PV module cleaning methods are broadly categorized as done in the following sections.

4.3.1 Forced flow from air-conditioning systems

A method proposed by Assi et al., mostly applied in the United Arab Emirates (UAE) and which could also be implemented in various other developed countries with a wide availability of air conditioners, uses airflow from the fan of air conditioners to force the flow onto solar panels for dust removal [11]. Fig. 4.1 is a conceptual diagram of such a system.

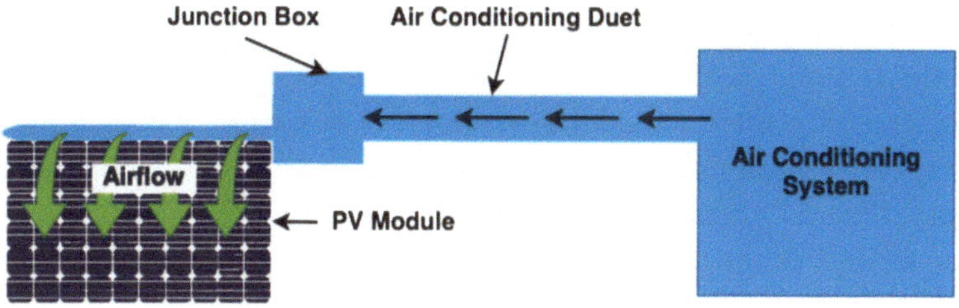

Figure 4.1 Conceptual diagram of forced air-flow-based cleaning.

4.3.2 Rainfall, water-based and manual cleaning

The rainwater incident on the tilted (for optimal solar irradiance-capture efficiency) solar panels cleans the loose dust from the PV surface. This process usually leaves behind the dust stuck due to moisture that generally requires heavier rainfall to remove it. The reliability of such a method to wash off the deposited soil is debatable when there exist intense soiling and inadequate rainfall. As reported by Kimber et al. (2006), a sharp decline in production was noticed following a light shower [12]. To avoid mud formation on PV panel surfaces and maintain optimal power generation, there is sometimes a need to avoid rainwater-based cleaning.

However, wind can be of assistance to reduce dust deposition. Although rains are not accurately predictable, the panel-surface property is improvised to increase the efficiency of this soiling-prevention method. Such improvisations will increase capital costs but would be useful in remote areas where maintenance is not always feasible. Some studies have revealed an average drop of less than 5% in energy production after dust buildup for a period of one year. Some may consider it to be a small price to pay compared to the cost incurred in cleaning the panels. However, during extended rainless periods, the reduction in power generation can exceed 20%.

One should not assume that solar panels in climates with frequent rainfall would need less maintenance. There's more to the equation than just dust. Pollen, bird droppings, leaves, ash, and a hazy film that can form during smog can combine into a layer of surface soiling, which may or may not wash away in the rain. In an overall economic sense, allowing rain to do the work of cleaning solar panels is a fine decision only in places where:

1. dust makes up most of the buildup that forms on the solar panels,
2. the makeup of the dust is such that rain will effectively wash it away,
3. it rains often enough to keep dust buildup from depleting the energy production of the solar panels.

With the wrong combination of pollutants, a little rain might not clean anything. It could just become the base ingredient that combines a collection of contaminants into a kind of slurry. Water-based cleaning requires water in large amounts, projected continuously at high pressure, to ward off any soiling particulate matter stuck on the PV panel surface.

Pressurized water mixed with a special cleansing agent helps clean off dust and cools down the PV panels installed in semi-arid regions and deserts. This method is a significant improvement over rainfall cleaning. However, there are many drawbacks to this method, such as a lack of reliability when applied in water-scarce regions, loss of water in huge quantities, possibility of chemical deposition on the PV panel edges, reduced efficiency as the pressurized water system consumes a significant amount of power, risks of water clogging, and need to refill water tanks (if any). Another factor is a fear of thermal shock in the hot PV panels because the water is more relaxed than the panel surface.

The potential for the attraction of more dust particles, when the PV panel surface is wet after cleaning, is also significant. These drawbacks are avoided by regulating the timing of cleaning to avoid thermal shocks and using power-efficient pumps to assure adequately pressurized water. Manual cleaning is analogous to the procedure one uses in the cleaning of high-rise building windows/glasses. Dust particles get scrubbed off using special brushes equipped with bristles to avert any scratches. Brushes connected with a direct water supply enable simultaneous scrubbing and washing. It is considerably more efficient than using rainfall to reinstate the solar panel surface to cleaner conditions. However, since the movement of brushes over the panel surface is uneven, a risk is always there that the direct contact of the brush with the PV panel may have an abrasive effect. It is also an expensive cleaning method because it requires skilled labor to clean off the soiling. The abrasion risk of PV panels can be reduced by using a soft cleaning cloth or a brush with soft bristles.

4.3.3 Mechanized cleaning and electrodynamic screens (EDS)

A set of mechanical apparatuses such as motors or robots must operate the brushes or wipers, and a water storage tank with sprinklers is used to clean the PV panel surface [13]. Mani et al. advocated cleaning the panel once a week in dry periods, but daily when the dust accumulation rate becomes intense [14]. This method uses automation where a micro-controller controls the system with the help of sensors. It is a reduced labor system for cleaning panel surfaces. However, it becomes beneficial under conditions when water-based cleaning is not feasible, and when the automated operation is essential. Due to its robust structure, it can also scare off birds. Although water can cool down the PV panels, the direct contact of such brushes and mechanized apparatus can be abrasive for cleaning panel surfaces. More energy consumption than any of the other already mentioned cleaning methods occurs, and maintenance of the mechanical parts is also needed. Its efficiency is uncertain, it may prove to be not quite efficient in case of high dust accumulation. A few measures are required to regulate the timing of cleaning to avoid thermal shocks, using power-efficient motors to operate brushes or wipers, and using softer cleaning materials to prevent abrasion of panel surfaces.

An electrodynamic screen (EDS) is a thin and transparent film that can be laminated on the top of solar panels. Such screens have the capability to move small particles when activated by a high-voltage signal. This means that that EDS placed on a solar panel can ensure automated and continuous removal of deposited dry dust without the use of water or any mechanical moving parts. It is an active dust-removal method that utilizes high-voltage supplies to power electrodes (to generate an electric field) of a transparent screen, removing the charged and uncharged dust particulate matter from the PV panel by shifting them to a panel surface edge. Over 90% of dust is detached within two minutes by this method, which proves to be useful in dry, arid, and desert regions [5]. This cleaning method is distinctively faster than others and consumes less power; a microcontroller controls the system with the help of sensors.

There is also a risk of screen degradation due to UV rays. The system also requires a high-voltage supply to generate an electric field, reducing the generation efficiency by 15%. It again proves ineffective for wet or cemented-dust particulates, so it is less efficient for small-sized particles. It is highly efficient for dry conditions and can be more effective against degradation of the screen by UV if weatherproof polymer or glass is used instead of the transparent screens for increased durability.

4.3.4 Super hydrophobic and hydrophilic planes (SHOP and SHIP)

Molecules that are attracted to, and tend to be dissolved by, water are called hydrophilic, and those that are not attracted to a mass of water and tend not to interact with water are called hydrophobic. Super-hydrophobic plane (SHOP) coatings completely repel water and heavy oils from any object. The desired item literally cannot be touched (or is repelled) by liquid and rolls off without touching the underlying surface, mitigating a myriad of possible issues. SHOP utilizes a tailor-made hydrophobic surface or chemical coating/screen layer that does not require any power whatsoever, so it is a passive cleaning method. The risk of screen degradation due to UV is also there but can be reduced by using an excellent weatherproof glass or coating. SHOP is moderately efficient when it rains but inefficient when it is dry. The layer does not allow dust to stick to the surface but needs water for dust removal. Thus, it becomes necessary to have rainfall or water to make this cleaning method effective. Some studies have proven that these modified PV panel surfaces tend to lessen the accumulation of soiling in atmospheric conditions [15]. On other surfaces for solar-power generation like concentrated solar-power reflectors, which are not possible to be effectively cleaned by other methods of planar surface cleaning like automated robotic devices, SHOP coatings are one of the few options.

Various studies aim at realizing super-hydrophobic surfaces by developing microstructures or nanostructures. With a hydrophobic coating, the layer forms a barrier. Water accumulates on the body in an almost spherical shape but gets blocked from adhering to this barrier's surface. When a treated surface tilts at an angle, the water rolls off like "a sphere rolling down a slide." Rain and water accumulated due to atmospheric condensation settle in the micro depressions present on the glass surface, attracting particulate matter. The water quickly evaporates, leaving residues that need removal. The nano-hydrophobic layer fills in the micro-depressions and provides a smooth glass surface that prevents water retention. However, it is required to verify the super-hydrophobic surfaces' applicability and feasibility in diverse environmental conditions [16]. One can determine that the nanocoated photovoltaic (PV) modules are comparatively cooler than the untreated ones through thermoelectric analysis. The hydrophobic coating layer reduces such issues and significantly reduces the menace of dust deposition on solar panels.

A Super-Hydrophilic Plane (SHIP) or surface uses nanofilms of titanium oxide, chemical coatings, and nano-patterned glass surface fabrication. This method utilizes a tailor-made hydrophilic surface or chemical coating/screen layer for soiling mitigation. In addition to the benefits like SHOP, this process can chemically break down the organic dirt through the reaction to UV light (photocatalytic effect) [17]. Thus, SHIP acts as a suspender material between dust and PV panel surface. It is more durable because it does not make use of a polymer-based coating, and so it can function for much longer periods than the SHOP surface [18]. However, more dust is accumulated when the coating deteriorates due to UV rays. It is moderately efficient in rain, but regular washing is required to wash off dust in dry climates.

He et al. found that the SHIP surface is inappropriate for solar-cell arrays in arid or dry regions [19]. However, in [20], the usage of TiO_2 film (photocatalytic) has been determined to be chemically stable, durable [21], non-toxic, economical, and transparent to visible light. Although most studies suggest the self-cleaning attribute of this surface mitigates the dust deposition, most of those have failed to provide any quantitative result on the restoration efficiency of PV panels, except for Hu et al. [22]. A summary of the various means of preventing the effects of soiling on solar panels that reduce the overall yield is presented [23]. With time, we expect more PV panels (and even parabolic reflectors outside the scope of this work) to be developed with self-cleaning abilities such that the requirement for external cleaning will be minimized.

4.4 Mechanisms behind degradation and techniques to detect degradation

Now, we'll look at how to monitor PV module degradation in the field. We can do this by analyzing the PV performance data and also by physically inspecting the modules. The performance ratio number that compares the module's actual power output to what we would expect given the irradiation and module temperature is useful for this purpose. If we see the performance ratio go down over time, it is because of factors like soiling. degradation, and faults in the modules themselves. In our approach, we measure the module power on the DC side of the inverter. So, differences in changes in the inverter won't come into play. Some degradation in a module's performance ratio over its lifetime is natural. There are microscopic cracks that can occur in the cells, resistances arise in the electrical paths, and, even on the atomic scale, there are undesirable changes in the cell.

Module design has dramatically improved, which reduces this degradation to a small amount. We tend to see in module warranties something like 90% warranted over the first ten years, then going down to 85% over 25 years. So, over the lifetime of the module, we're looking at a degradation rate of less than 1%. Now, a 1% change in performance ratio is a minimal change. Since the value is subject to a lot of noise, like soiling, solar spectrum, and differences in radiation measurements at different places on the site, we shouldn't be too concerned if we occasionally see a number greater than this. Remember, the primary use of performance ratio is to detect significant faults in our PV modules over time.

We also mentioned that physical inspections could be useful. You would think that any visible fault would show up in the PR data, but actually, we have seen at test sites modules with cracked glass and delamination in which the PR was in the normal range. So we do need to do visual inspections. Things to look out for are glass breakage, snail-trails in the cells, yellowing encapsulant, cracking of the back-sheets, burn marks at the interconnects, and discoloration around the edges of frameless modules where humidity has seeped in. A convenient tool for field inspections is an infrared camera, which shows where the hot-spots in cells appear by cell over-heating, can damage the module, and even pose a safety risk, so it's worth spotting.

Manufacturers of PV Module provide, currently, a power guarantee for 20–25 yr. However, one can ask: How reliable are the PV Modules in the field? To date, we do not have enough statistical data from the field to answer this question. Even the oldest PV installations still have yet to reach that 25-yr lifetime. The answer to this question is critical because it affects the Levelized Cost of Energy produced by the PV technology. For that purpose, the performance and the reliability of PV modules and their components need assessment based on well-defined tests. Standard test designs assess the performance, degradation rates, and the effect of specific external factors under prolonged exposure to typical climates. However, there are no formal tests for desert climates yet. That would require a proper assessment of the effect of high-temperature cycling, humidity, high ultraviolet solar radiation, and module material resistance to desert climate conditions. Standard tests define the minimum requirements. They help identify design flaws and causes for module failures to address safety qualifications.

General requirements are unified for all technologies, while the testing methods are specific to silicon and thin-film technologies, respectively. According to IEC's 61215 standards, PV modules are tested for crystalline silicon and IEC's 61646 for thin films. The performance assessment, by measuring the maximum power, P_{max}, at standard conditions, are the temperature coefficient, nominal operation module temperature, the insulation resistance, the wet leakage current, thermal tests for the bypass diode, the outdoor exposure to UV, temperature cycles, and humidity freeze—and, finally, the mechanical tests to evaluate the robustness of terminations and resistance to hail impact.

In summary, the necessary modifications, especially in arid and desert locations include:

1. Additional thermal cycling to: extend the test temperate including bypass diodes.
2. UV exposure: increase the UV exposure time and damp-heat tests.
3. Optical modifications.
   - Sand abrasions: ensure that the module glass and backsheet are unaffected by sandstorms.
   - Ammonia corrosion testing.
   - Dynamic mechanical loading.
   - Potential Induced Degradation (PID), which will be discussed in great detail in subsequent chapters [24].
   - Failure mechanisms missed by IEC 61215: PID and cracked cells.

In the next chapter, we compare the results of Hu et al. with the actual energy-yield increase obtained through our automated soiling mitigation system in our institute's rooftop solar-panel installation. We also present an indigenous development of an automated cleaning system intended to mitigate the effect of dust deposition.

Bibliography

[1] V. Gupta, M. Sharma, R.K. Pachauri, K.D. Babu, Comprehensive review on effect of dust on solar photovoltaic system and mitigation techniques, Solar Energy 191 (2019) 596–622, https://doi.org/10.1016/j.solener.2019.08.079.
[2] Y.N. Chanchangi, A. Ghosh, S. Sundaram, T.K. Mallick, Dust and PV performance in Nigeria: a review, Renewable & Sustainable Energy Reviews 121 (2020) 109704, https://doi.org/10.1016/j.rser.2020.109704.
[3] Energy from the desert: practical proposals for very large scale photovoltaic systems - summary.
[4] D.Y. Goswami, Principles of Solar Engineering, third edition, CRC Press, Taylor & Francis Group, Boca Raton London New York, 2015, oCLC: 913486518.
[5] R. Kohli, K.L. Mittal, Developments in Surface Contamination and Cleaning, Volume 3, 2011.
[6] D. Goossens, Z. Offer, A. Zangvil, Wind tunnel experiments and field investigations of eolian dust deposition on photovoltaic solar collectors, Solar Energy 50 (1) (1993) 75–84, https://doi.org/10.1016/0038-092x(93)90009-d.
[7] E. Aguado, Effect of advected pollutants on solar radiation attenuation: Mojave Desert, California, Atmospheric Environment. Part B, Urban Atmosphere 24 (1) (1990) 153–157, https://doi.org/10.1016/0957-1272(90)90020-u.
[8] H. Kawamoto, T. Shibata, Electrostatic cleaning system for removal of sand from solar panels, Journal of Electrostatics 73 (2015) 65–70, https://doi.org/10.1016/j.elstat.2014.10.011.
[9] T. Sarver, A. Al-Qaraghuli, L.L. Kazmerski, A comprehensive review of the impact of dust on the use of solar energy: history, investigations, results, literature, and mitigation approaches, Renewable & Sustainable Energy Reviews 22 (2013) 698–733, https://doi.org/10.1016/j.rser.2012.12.065.

[10] M.R. Maghami, H. Hizam, C. Gomes, M.A. Radzi, M.I. Rezadad, S. Hajighorbani, Power loss due to soiling on solar panel: a review, Renewable & Sustainable Energy Reviews 59 (2016) 1307–1316, https://doi.org/10.1016/j.rser.2016.01.044.
[11] A. Assi, A. Hassan, M. Al-Shamisi, H. Hejase, Removal of air blown dust from photovoltaic arrays using forced air flow of return air from air conditioning systems, in: 2012 International Conference on Renewable Energies for Developing Countries (REDEC), IEEE, 2012.
[12] A. Kimber, L. Mitchell, S. Nogradi, H. Wenger, The effect of soiling on large grid-connected photovoltaic systems in California and the southwest region of the United States, in: 2006 IEEE 4th World Conference on Photovoltaic Energy Conference, IEEE, 2006.
[13] D.G. für Sonnenenergie, Planning and Installing Photovoltaic Systems, 2007.
[14] M. Mani, R. Pillai, Impact of dust on solar photovoltaic (PV) performance: research status, challenges and recommendations, Renewable & Sustainable Energy Reviews 14 (9) (2010) 3124–3131, https://doi.org/10.1016/j.rser.2010.07.065.
[15] M. Jaszczur, J. Teneta, K. Styszko, Q. Hassan, P. Burzyńska, E. Marcinek, N. Łopian, The field experiments and model of the natural dust deposition effects on photovoltaic module efficiency, Environmental Science and Pollution Research 26 (9) (2018) 8402–8417, https://doi.org/10.1007/s11356-018-1970-x.
[16] Y.-B. Park, H. Im, M. Im, Y.-K. Choi, Self-cleaning effect of highly water-repellent microshell structures for solar cell applications, Journal of Materials Chemistry 21 (3) (2011) 633–636, https://doi.org/10.1039/c0jm02463e.
[17] W. Thongsuwan, T. Kumpika, P. Singjai, Effect of high roughness on a long aging time of superhydrophilic TiO2 nanoparticle thin films, Current Applied Physics 11 (5) (2011) 1237–1242, https://doi.org/10.1016/j.cap.2011.03.002.
[18] L. Verma, M. Sakhuja, J. Son, A. Danner, H. Yang, H. Zeng, C. Bhatia, Self-cleaning and antireflective packaging glass for solar modules, Renewable Energy 36 (9) (2011) 2489–2493, https://doi.org/10.1016/j.renene.2011.02.017.
[19] G. He, C. Zhou, Z. Li, Review of self-cleaning method for solar cell array, Procedia Engineering 16 (2011) 640–645, https://doi.org/10.1016/j.proeng.2011.08.1135.
[20] K. Nakata, A. Fujishima, TiO_2 photocatalysis: design and applications, Journal of Photochemistry and Photobiology. C, Photochemistry Reviews 13 (3) (2012) 169–189, https://doi.org/10.1016/j.jphotochemrev.2012.06.001.
[21] L. Graziani, E. Quagliarini, F. Bondioli, M. D'Orazio, Durability of self-cleaning TiO_2 coatings on fired clay brick façades: effects of UV exposure and wet & dry cycles, Building and Environment 71 (2014) 193–203, https://doi.org/10.1016/j.buildenv.2013.10.005.
[22] J. Hu, N. Bodard, O. Sari, S. Riffat, CFD simulation and validation of self-cleaning on solar panel surfaces with superhydrophilic coating, Future Cities and Environment 1 (2015) 8, https://doi.org/10.1186/s40984-015-0006-7.
[23] D. Deb, N.L. Brahmbhatt, Review of yield increase of solar panels through soiling prevention, and a proposed water-free automated cleaning solution, Renewable & Sustainable Energy Reviews 82 (2018) 3306–3313, https://doi.org/10.1016/j.rser.2017.10.014.
[24] S. Mishra, K. Bhargava, D. Deb, Numerical simulation of potential induced degradation (PID) in different thin-film solar cells using SCAPS-1d, Solar Energy 188 (2019) 353–360, https://doi.org/10.1016/j.solener.2019.05.077.

CHAPTER 5

Water-free automated solar-panel cleaning

In this chapter, a detailed review of soiling prevention methods is presented to increase solar panels' energy capture. We describe both semiautomatic water-based solutions and automatic water-free cleaning solutions. The technical details of an indigenously developed automated, water-free cleaning device to remove soiling of the solar panels are discussed. We compare the energy-capture levels, over the course of a month, for PV panels regularly cleaned using automated solar panel cleaning solution, with that of soiled panels in a local solar-panel installation. The increased yield is measured. One of the most important aspects of the solar-power system, for example, in rooftop solar systems, is the cleaning of the solar panels. There are significant losses due to dirt and soiling of the solar panels. So, it becomes very important to keep the solar panels clean.

5.1 Semiautomatic water-based cleaning solutions

Because of shadows from neighboring buildings or trees and other obstructions, in the residential sector, there can be situations when one may have to elevate the structure hosting the panels which can then be cleaned automatically by a pumping arrangement running through the motor to the solar panels.

The pipes are anchored to the channel, using butterfly clamps to hold the anchor rod on which we place the clamp to hold the pipes. For this water-based cleaning mechanism, which is the most common mechanism to date, a pump and a pump controller are used to suck the water from the tank. The efficiency of the pump is an important factor in this arrangement to ensure that energy losses are minimal during cleaning. The controller is programmed to switch on for 3–5 minutes on about 3–4 days a week, for a regular and individually owned rooftop arrangement [1]. The frequency of cleaning can be increased for larger installations and in more dust-prone areas [1].

The efficiency of this sort of cleaning solution is, at best, medium, and there is a need to schedule the cleaning action to ensure that there is no

Degradation, Mitigation, and Forecasting Approaches in Thin Film Photovoltaics
https://doi.org/10.1016/B978-0-12-823483-9.00016-4

thermal damage. The best time to clean the panels with water is early mornings or evenings when the body temperature of the modules is low enough so that the toughened glass of the module does not break due to the temperature difference between the water and the heated surface. The water-based cleaning mechanism cools down the PV system [2] and improves on the rainfall cleaning method described in the previous chapter. Also, the air's moisture content aids in cleaning in the mornings, and so the usage of water is less [3]. The preferred pH of the pressurized water is between 6–8 to avoid any corrosion on the panel surface. The temperature of the water should be less than 25°C [4].

To mechanize the entire process, an apparatus like a motor can also be used to operate the brush and/or wiper to control the water flow [5]. In such systems, the panels can be cooled with water, and the automated portion of the operation is supplemented with sensors and controllers [6]. There is reduced labor cost with such a semiautomated mechanism. Some of the drawbacks are that direct contact with a panel surface is abrasive, some energy is consumed in the cleaning process, and higher capital and periodic maintenance costs are incurred.

However, water is a scarce resource, and at the same time it can be sometimes counter-productive when cleaning may worsen the condition of dust deposition by making it stick even more to the surface. Water-based automated cleaning is ineffective in dry areas. Such methods typically warrant a large amount of water usage and often results in chemical deposition and water clogging, thereby resulting in reduced power output. These methods also require water-tank refilling, the dust can act like a magnet when wet, and there is always a fear of thermal shock to the solar modules. Therefore, there is a need to explore water-free automated solutions that are cost-effective and customized as per the dust characteristics of the location [7].

Next, we present an indigenously developed water-free and automated solar-panel cleaning robot with minimum viability of commercialization. How safe is it to operate a waterless solar-panel cleaning unit, and how about scratches on PV panels? The proposed solar-panel cleaning-robot development involves stretchable, scratch-proof microfiber cloth, thereby making it safe against scratches, and protects the anti-reflective coating of the panels with a surface active agent [8].

5.2 Automated panel cleaning unit

Costa et al. have discussed many publications relating to soiling prevention, inspection, and various ways to clean PV panels [9]. Sarver et al. have suggested several steps for dust mitigation such as: (1) preventive, durable and non-toxic coatings for various conditions of the dusty environment may be applied (or reinforced) either during actual use or at the time of manufacture; (2) automated mechanical cleaning systems (air, water, solution) for large collector fields, including: (a) waterless or water-conserving (recycling) and (b) mobile operated or attached to the arrays; and (3) sensors to detect the critical point for cleaning of surfaces (i.e., to initiate automatic cleaning techniques or alert about problems with loss of dust-coating effectiveness) [10].

There is a huge variation of soiling patterns from one location to another, rendering such a prediction difficult. To accurately predict the soiling, a broader database of every climatic zone helps to design an algorithm that produces a feasible soiling mitigation solution ideally suited to the desired location by approximately forecasting the increased yield. Such a process also involves a forecast-based calculation for the prediction of the break-even in an installed solar farm. For an accurate prediction of location-specific soiling patterns and losses incurred in the generation, these aspects can be identified:

1. the physical properties and dust characterization;
2. a thorough study for minimal dust accumulation across multiple configurations of PV modules that considers aspects like local wind patterns;
3. scheduling an optimal PV system-maintenance cycle considering prevailing issues of the climate and environment.

Some available autonomous solutions are inefficient due to their heavy weight and uneconomical services, thus reducing their practical commercial viability. We have proposed and designed a water-free cleaning solution for solar panels that is newly preprogrammed, efficient, economical, and eco-friendly. It combines several mechanisms controlled through various electronic components, thus performing mitigated and optimized solar panel cleaning. The proposed solution is service effective (optimized cleaning), environmentally friendly (water-free), energy independent (no external power requirement), self-automated (labor-free), long lasting, and cost-effective.

India has set a new solar target of 100 GW to be achieved by 2022. The issue of tackling climate change is possible with solar power, but the increasing water scarcity is a significant threat to solar expansion in India.

The energy capacity in the water-scare areas of India is expected to increase by 79%. The issue intensifies when it comes to solar panels in arid and desert regions. We estimate that the water required for the operation and maintenance of solar panels in India would lie between 7000 and 200, 000 l/MW per wash, considering a weekly cleaning frequency. If the water constraints are factored in, the actual proportion of India's power produced by solar-power plants will be much lower than now estimated. While other solar infrastructure issues may need addressing, an automated solution based on 100% water-free solar panel-cleaning technology is necessary to improve energy yield and ensure a sustainable, cost-effective maintenance solution.

The cleaning apparatus senses the current status and manages the panels' degree vis-a-vis user-defined data. The solar rows are inclined so that the upper end is positioned at a specific angle (for maximal irradiance) to the horizontal plane. This cleaning system consists of a frame structure containing the cleaning apparatus and a drive system to operate laterally and selectively to clean longitudinally. The complete system moves in either direction, as needed, with the initiation of the cleaning operation. The ends of the solar rows come with two engagement systems to (i) lock and dock the system and handle charging operation; and (ii) for cleaning and resting of the apparatus.

Fig. 5.1 shows an indigenously built automated cleaning unit for sustainable operations with minimal capital and maintenance expenditures. It works automatically based on an optimized technique for managing the power delivered to the system, and it provides high-quality cleaning with the use of precipitators, which accumulate the dust present on the brush.

Fig. 5.2 shows an isometric view of the service unit with the cleaning brush supported by a combination of stainless-steel rods and metallic boxes at each brush edge.

These boxes are outside the upper and lower boundaries of the solar row. The automated service unit is lightweight and performs cleaning in the lateral direction (parallel to panels). The brushes are placed in the longitudinal direction to complete the cleaning task in the least possible time. With no usage of guide rails, water, or any other liquid, it is easy to maintain, and the functioning of the apparatus requires minimal fixed cost. The service unit is automatically charged and functions without water or liquid for the entire cleaning operation. The system can also adjust the sensor's angle for uninterrupted motion with solar panels integrated into the solar tracking system. User-defined data controls the movement, thereby punc-

Figure 5.1 Automated solar panel-cleaning service unit.

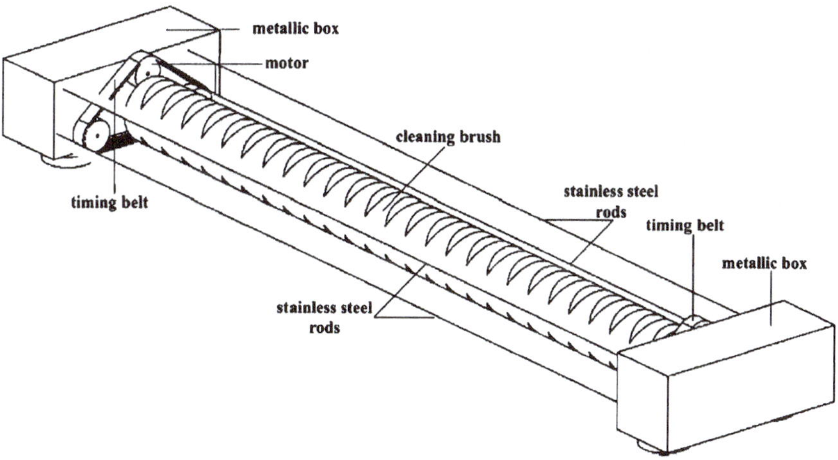

Figure 5.2 Isometric view of service unit for water-free automated cleaning.

tuating the unit's direction and range. The designed system works on the panels' edge, thereby eliminating the need for guide rails.

The service unit traverses normal to the cleaning brush and along the solar row through a crawler system containing three wheels positioned in an appropriate triangular pattern. The wheel at the apex gets affixed to a motor that rotates the wheels through the conveyor belt. The conveyor belt acts as an interface between the service unit and the solar-panel surface. The friction due to the conveyor belt with the glass surface causes the unit's displacement in either direction. The brush position is such that there is maximal cleaning efficiency, and there is no damage to the surface of the solar panels and cleaning brush due to friction. For better maintenance of the cleaning brush, a characteristic feature such as longitudinal extension for changing the brush position is also available. The metallic boxes mainly protect all the electronics, including the microcontroller, battery housing, battery, motor drivers, etc., from external disturbances such as sudden jerks and unfavorable weather conditions, if any.

5.3 Specifications of service unit

Essential items used in the automated service unit and their configurations are:

1. Arduino is an open-source physical computing platform based on a simple i/o board and a development environment. Arduino is useful for developing stand-alone interactive objects or connections to the software loaded on Flash, Processing, and MaxMSP computing units. The open-source IDE is downloadable for free on platforms like Mac OS X, Windows, and Linux. Arduino UNO is a user-friendly and multipurpose board with the microcontroller ATmega328P (Input voltage: 6–20 V; length (68.6 mm), width (53.4 mm), clock speed: 16 MHz), used for intelligent control. It comprises six analog inputs, 14 digital input/output pins (*six* are usable as Pulse Width Modulation (PWM) outputs), a USB connector, a power jack, an in-circuit serial programming (ICSP) header, and a reset button [11].
2. Dual DC-Motor Driver (RKI-1341) is ideal for this application to deliver 10 A current. It has a simple TTL/CMOS-based interface for direct connection to ATmega328P's i/o ports. It has a braking feature to guarantee an immediate halt of the motors' shaft, which is appropriate for this application and does not negatively affect MCU performance. This motor driver drives a DC geared motor (Model no. RKI-1418 60 RPM).

3. L298N Dual Motor Driver is a high-voltage and high-current dual motor driver designed to accept logic commands to drive relays, solenoids, DC motors, stepper motors, etc. The two enabled inputs function as an enable/disable switch. This motor driver drives a DC geared motor (Model no. RKI-1188 300 RPM).
4. HC-05 Bluetooth Module is an additional feature facilitating the service unit through an Android application. This module is configurable for baud rates ranging from 1.200 to 11.5200 bps to avoid microwave interferences, if any. The module communicates with the help of USART (Universal Synchronous/Asynchronous Receiver/Transmitter) and easily interfaces with any microcontroller supporting USART. We also configure the default settings by using the command mode. So, when looking for a Wireless module that could transfer data from the computer or mobile phone to a microcontroller or vice versa, this module is the right choice.

A detailed description of all these items and their associated datasheets are provided [12]. Many researchers are also investigating options to make solar panel-cleaning units self-dependent in terms of energy by providing a solar module of the size of the cleaning robot and placing it on top. Such a mechanism can ensure that the unit itself does not need significant additional or separate charging-related downtime when trying to service a reasonably large solar farm.

5.4 Process command flow of operations in a MicroController

Fig. 5.3 shows the flow of process commands of the water-free automated cleaning system. The service unit hosts independent circuits that control the service unit's primary movement along the solar row, in either direction, normal to the unit's longitudinal axis. The circuit's primary function is to define the unit's start and end position using multiple sensors to measure the distance covered, which is fed as a control input to process the status in terms of any further movement of the unit. The RKI-1418 DC geared motor is responsible for the service-unit motion, and its driver (L298N) gets feedback from the microcontroller. The circuit contains a detector (Sensor-1) to measure the state of battery charge, which controls the service unit's further movement. Using initial user-defined inputs, we predetermine the required battery level for the service unit's efficient functioning. Battery level and distance covered are further compared with the user-defined data

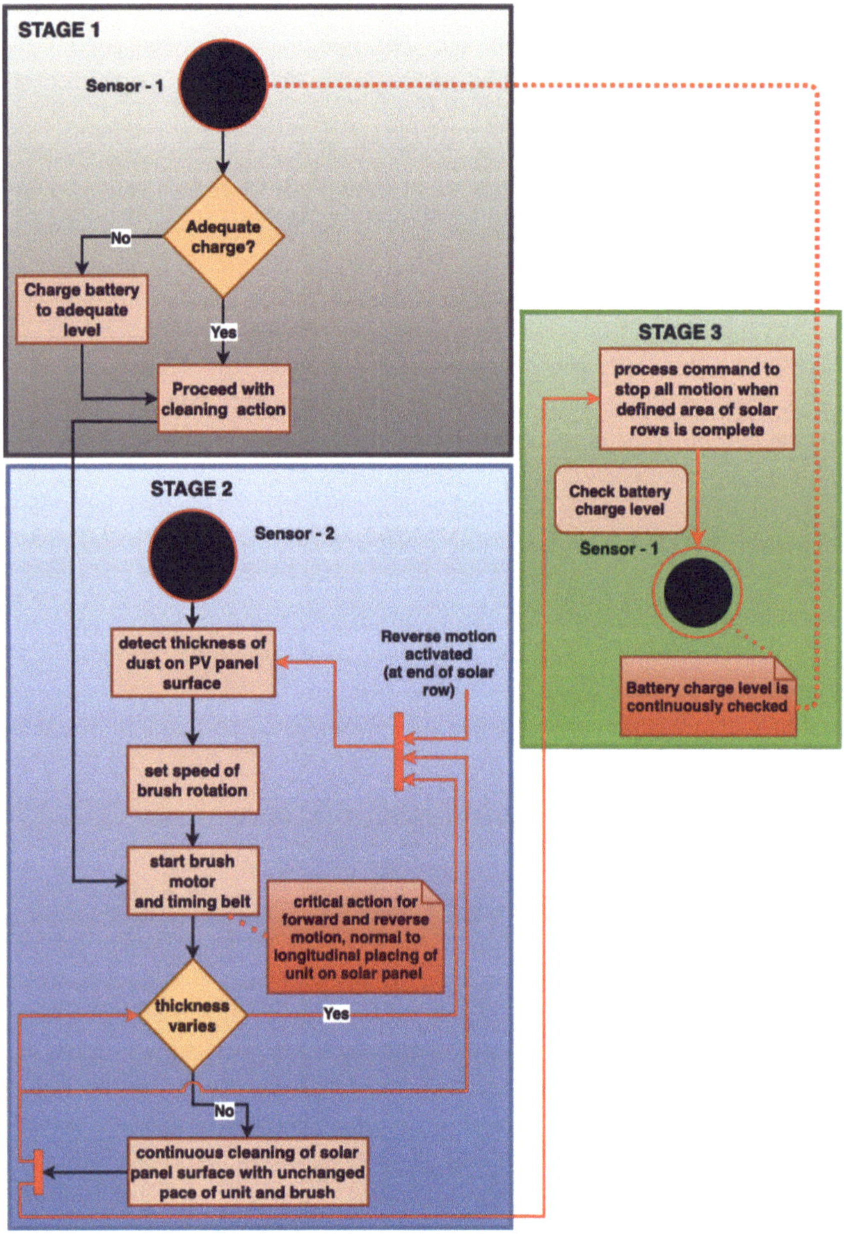

Figure 5.3 Process command flow of service unit's microcontroller operation.

(as per the demand) to ensure further cleaning of the solar row. The circuit configuration for net-metering facilitates providing surplus electricity to the grid when the battery has adequate charge.

The circuit of the Sensor-2 assembly measures dust accumulation in terms of the layer thickness to set the speed variation of the unit or the input to the L298N motor-driver driving RKI-1418 DC geared motors. Suppose the dust-layer thickness is greater than the user-defined data. The microcontroller may decrease the unit's speed from the normal levels by varying the timing belt's speed for increased revolutions of the brush in a controlled area. The RKI-1188 DC geared motor is responsible to rotate the cleaning brush, and its driver (RKI-1341) receives feedback from the microcontroller concerning the level of dust accumulation.

The instant the unit reaches the second engagement section at the other end of the solar row is detected by an edge-detection sensor (HC-SR04 Ultrasonic ranging module allows 2–400 cm non-contact measurement with a ranging accuracy of 3 mm) when an obstacle berates the sensor. Brushes get cleaned by rubbing against the electrostatic precipitators to collect the accumulated particles. Once cleaned, the unit retraces the track under the control of the edge-detection sensor.

The data acquired by Sensor-2 is compared with the user-defined data, which accordingly sets the speed of the service unit. The brush position can also be altered suitably as per dust-layer thickness, and this reduces necessary brush maintenance.

5.5 Quantitative measurement of the effect of cleaning

A minimum viable product (prototype), as shown in Fig. 5.1 and Fig. 5.2, was built indigenously and cost about $450.00. With mass production, we expect that the cost of this automated service unit will go down steadily. We set up this solar panel-cleaning solution in the rooftop installation of solar panels at the Institute of Infrastructure Technology Research and Management, Ahmedabad, India. Each of the two 60 kW inverters from KACO Inc., connected to 820 solar modules, was used.

We reasonably assume that the two sets of modules' solar irradiance are the same over a month since the panels connected to Inverter-1 and Inverter-2, respectively, are atop the same terrace. We clean the first set of soiled panels connected to Inverter-1 and the second set to Inverter-2 once daily, starting at 8 a.m. (before peak solar irradiance). The total energy supplied by these panels in May 2017 (before the onset of monsoonal rains)

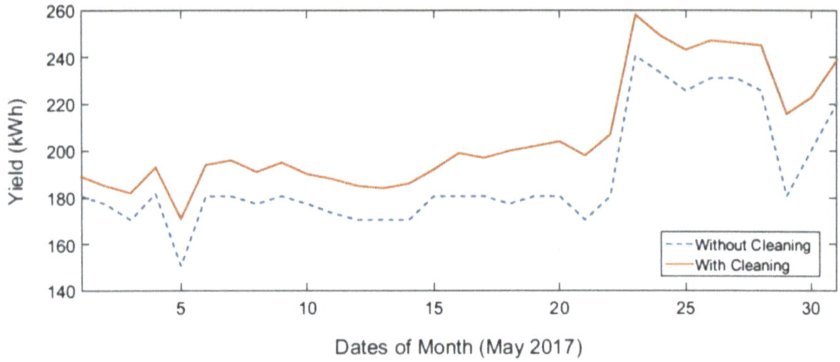

Figure 5.4 Comparison of yields from root-top solar panels, with and without an automated cleaning system.

provides the yield data (kWh). Fig. 5.4 shows that the energy capture is consistently higher, as measured through Inverter-2 with daily cleaning, compared with the readings reported through Inverter-1 with no cleaning, thereby indicating the effectiveness.

There is an increase of 35.4 kWh on May 29, which could be an aberration. However, over the entire month, there is an increase of 9.05% in kWh with cleaning, as shown in Table 5.1.

Table 5.1 Comparison of energy captured with and without automated cleaning.

Energy captured (in kWh)		Condition of panels	Efficiency increase
Total energy	5864	uncoated; without artificial cleaning	
Mean daily energy	189.16		
Minimum (daily)	150.8		
Maximum (daily)	240.8		
Total Energy	6394	uncoated; with daily cleaning	9.05% (over entire month)
Mean daily energy	206.26	using automated device	
Minimum (daily)	171		
Maximum (daily)	258		

Considering the cost of the cleaning unit, the effective increase is about 6.31%, which is still quite significant. Suppose we assume that the service unit's life is about ten years, incurring a daily cost of $0.5 (the equivalent of energy), to complete the entire cleaning operation. In that case, there is an effective cost of $0.62 daily after taking into account the depreciation of the cleaning device. With an assumed cost of about 12 cents for 1 kWh, there is a 5.16-kWh loss of useful energy daily, that is, 160 kWh in May 2017.

Therefore, the adequate total energy with cleaning is only 6234 (= 6394–160) kWh, which is an effective increase of 6.31% over the uncleaned condition (5864 kWh). We account for the device's purchase cost and the battery-charging cost (equivalent to the power consumed) in these calculations. When compared with the typical value of approximately 3.6–5.4% increase in efficiency given by the SHOP solution suggested by Hu et al. [13], this proposed solution with a rise in 6.31% efficiency is very encouraging. With bulk manufacturing, it is possible to approach 9.05%.

Bibliography

[1] S. Krauter, Increased electrical yield via water flow over the front of photovoltaic panels, Solar Energy Materials and Solar Cells 82 (1–2) (2004) 131–137, https://doi.org/10.1016/j.solmat.2004.01.011.

[2] H. Bahaidarah, A. Subhan, P. Gandhidasan, S. Rehman, Performance evaluation of a PV (photovoltaic) module by back surface water cooling for hot climatic conditions, Energy 59 (2013) 445–453, https://doi.org/10.1016/j.energy.2013.07.050.

[3] S. Odeh, M. Behnia, Improving photovoltaic module efficiency using water cooling, Heat Transfer Engineering 30 (6) (2009) 499–505, https://doi.org/10.1080/01457630802529214.

[4] D.-J. Kim, D.H. Kim, S. Bhattarai, J.-H. Oh, Simulation and model validation of the surface cooling system for improving the power of a photovoltaic module, Journal of Solar Energy Engineering 133 (4) (2011), https://doi.org/10.1115/1.4004508.

[5] L.A. Lamont, L.E. Chaar, Enhancement of a stand-alone photovoltaic system's performance: reduction of soft and hard shading, Renewable Energy 36 (4) (2011) 1306–1310, https://doi.org/10.1016/j.renene.2010.09.018.

[6] Z. Yuyi, Z. Yu, L. Huanxin, L. Yunjia, L. Liang, Control system design for a surface cleaning robot, International Journal of Advanced Robotic Systems 10 (5) (2013) 220, https://doi.org/10.5772/56200.

[7] J.J. Zhen, Z.M. Yang, Design and realization of a flat surface cleaning robot, Advanced Materials Research 505 (2012) 477–483, https://doi.org/10.4028/www.scientific.net/amr.505.477.

[8] K. Moharram, M. Abd-Elhady, H. Kandil, H. El-Sherif, Influence of cleaning using water and surfactants on the performance of photovoltaic panels, Energy Conversion and Management 68 (2013) 266–272, https://doi.org/10.1016/j.enconman.2013.01.022.

[9] S.C. Costa, A.S.A. Diniz, L.L. Kazmerski, Dust and soiling issues and impacts relating to solar energy systems: literature review update for 2012–2015, Renewable & Sustainable Energy Reviews 63 (2016) 33–61, https://doi.org/10.1016/j.rser.2016.04.059.

[10] T. Sarver, A. Al-Qaraghuli, L.L. Kazmerski, A comprehensive review of the impact of dust on the use of solar energy: history, investigations, results, literature, and mitigation approaches, Renewable & Sustainable Energy Reviews 22 (2013) 698–733, https://doi.org/10.1016/j.rser.2012.12.065.

[11] J. Henderson, J. Condell, J. Connolly, D. Kelly, K. Curran, Review of wearable sensor-based health monitoring glove devices for rheumatoid arthritis, Sensors 21 (5) (2021) 1576, https://doi.org/10.3390/s21051576.
[12] D. Deb, N.L. Brahmbhatt, Review of yield increase of solar panels through soiling prevention, and a proposed water-free automated cleaning solution, Renewable & Sustainable Energy Reviews 82 (2018) 3306–3313, https://doi.org/10.1016/j.rser.2017.10.014.
[13] J. Hu, N. Bodard, O. Sari, S. Riffat, CFD simulation and validation of self-cleaning on solar panel surfaces with superhydrophilic coating, Future Cities and Environment 1 (2015) 8, https://doi.org/10.1186/s40984-015-0006-7.

CHAPTER 6

Numerical simulations of potential-induced degradation

The leakage current produced in TFPV cells leads to potential-induced degradation at the cell level. Various experimental methods are generally employed to investigate the progression of potential-induced degradation in cells and modules. However, these methods are quite cumbersome and cause wastages of time and expensive resources. In this chapter, the simulation-based analysis investigates PID progression in TFPV cell-based variety of absorber layers. Furthermore, the analysis is presented in terms of the rising levels of PID in simulated cells. The analysis yields an important observation that the PID effect in TFPVs is dependent on the absorber layers' properties. The presented analysis is useful for the assessment and prediction of the PID effect at the cell level.

6.1 Basics of potential-induced degradation

Generally, TFPV modules suffer performance degradation due to high-voltage stress generated from the grounded frames. These stresses are proportional to the leakage current flowing between the TFPV cell-matrix and ground. The repetitive application of these stresses leads to potential-induced degradation (PID), causing a severe deterioration in open-circuit voltage (V_{OC}) and, subsequently, power-conversion efficiency. Alternatively, one can understand the deteriorating effect of PID in terms of reduced shunt resistance and increased recombination in the depletion region. In crystalline silicon modules, decreased shunt resistance causes a reduction in both the open-circuit voltage and fill-factor. PID shunts' progression increases with the diffusion of sodium ions (Na^{+}) inside the cell structure, and its origin is soda-lime glass substrates used for fabrication [1].

Several module-level tests are normally used to verify the PID induced degradation [2]. Studies in the past have revealed that PID could also become evident shortly after installation (maybe after a week or so) [3]. Modeling the phenomena provides an opportunity to dig deeper into the concept of PID occurrence in TFPVs. The mathematical analysis of changes happening in PV-cell electrical parameters has revealed that the

Degradation, Mitigation, and Forecasting Approaches in Thin Film Photovoltaics
https://doi.org/10.1016/B978-0-12-823483-9.00017-6

value of the ideality factor (η) increases with the progression of PID due to the generation of electronic defect states attributed to the movement of Na^+ ions. Steingrube's model explained the reason for an increment in the value of η by showcasing its dependence on coupled defect levels both theoretically and experimentally [4]. In particular, the value of η becomes greater than 2 with the onset of the PID effect inside the TFPV cells because, with the ingress of Na^+ ions, both shunt and series resistance changes [5].

The effect of PID has also been observed to be affected by material properties. Thus, distinct PID progression levels are expected in various TFPV modules, although minimal exploration has been done in this direction [6]. Generally, the CIGS TFPVs are fabricated on soda-lime glass substrates. The PID effect is frequently observed in these modules shortly after installation, which is attributed to the modified charge-carrier dynamics inside the cell structure [7]. PID is related to creating crystallographic stacking faults inside various semiconductor layers inside the TFPV cell structure. These faults especially produce defect states in the absorber layer's band gap, thereby causing shunting of the p–n junction. The other relevant details of PID in TFPV cells and modules have already been described in Chapter 3.

Several experimental techniques, such as electroluminescence imaging and thermal imaging, among others, can be employed to characterize the PID progression inside TFPV cells and modules. However, these experimental investigations are quite cumbersome. Instead, modeling can offer a more convenient way to assess PID progression in cells and modules. In the literature, a minimal number of reports are available that are dedicated to the studies about PID progression in TFPV cells [8]. Various licensed and free software is available for the analysis of solar PV cells. The solar-cell capacitance simulator (SCAPS) is a one-dimensional freely available software, developed at the University of Gent, used for simulating the thin-film PV structures [9].

Simulation-based modeling of TFPV devices has proved to be quite useful and accurate in predicting the performance improvement and degradation of TFPV cells. This chapter discusses the simulations about the effect of PID in TFPVs based on several types of absorber-layer materials. The effect of PID progression in several types of TFPV cells is mapped in terms of varying defect concentrations at different locations inside the cell structure. The analysis highlights the importance of countering the PID progressions inside cells to increase the TFPV modules' reliability through mathematical modeling. The analysis also compares the impact of absorber-layer properties on PID progression in cells.

6.2 Simulation methodology

The simulations about the effect of PID in various types of TFPV cells are performed using the TFPV simulation tool SCAPS-1D (version 3.3.05). In particular, this tool is suitable for doing the interfacial and bulk analysis of cells [10]. The tool is freely available (although permission is required from the inventor to publish the results obtained through this software) to researchers working in TFPV modeling and simulation and can run on computers having the Windows operating system with minimal hardware requirements. SCAPS, initially developed for simulating the CdTe and CIGS TFPVs, has also demonstrated the capability to simulate advanced TFPV cells based on organic and perovskite absorber materials. Some exclusive features of SCAPS are:

- Up to seven semiconductor layers can be defined
- Capable of grading of material parameters
- Includes direct, Auger, and SRH recombination mechanisms
- Capable of performance analysis in terms of voltage, frequency, and temperature
- The device can be illuminated from either side
- Output is generated in terms of current-density voltage, quantum efficiency-wavelength, and energy band-position characteristics
- Batch simulation of parameters is possible
- Interactive user interface
- All parameters, such as band gap, electron affinity, states' density, dielectric constant, carrier mobility, doping concentration, and the absorption coefficient, can be graded
- The effect of bulk and interfacial defect states can be analyzed.

To replicate a scenario similar to PID progression, defect states have been selectively defined inside the model. These defect states are defined as being close to the middle of the band gap (deep traps). As it is essential to ensure the correctness of the adopted model, which relies on the accuracy of the incorporated simulation parameters, all the electrical and optical parameters in the simulation domain have been adopted from a reliable source [11].

The graphical user interface (GUI) of SCAPS appears in Fig. 6.1.

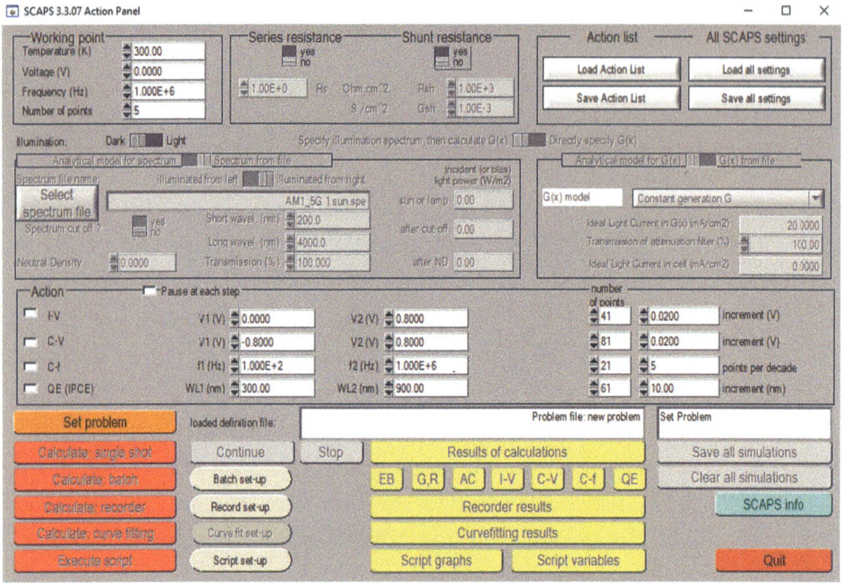

Figure 6.1 The graphical user interface SCAPS-1D.

SCAPS can be utilized to model TFPVs very efficiently under various circumstances produced by absorption profiles, material properties, defects and recombinations, doping concentrations, tunneling, grading, temperature, etc. An essential feature of SCAPS is that almost every photovoltaic material can be incorporated, such as organic, perovskite, and quantum dots. Fig. 6.2 shows an overview of a sample model definition inside SCAPS for perovskite TFPV.

The numerical settings are crucial for obtaining accurate SCAPS results; therefore, special care is needed after the model creation. A typical numerical settings panel inside SCAPS appears as Fig. 6.3.

The working temperature is set at 300 K, and the series and shunt resistance values are maintained as defaults. The structure is illuminated from the side of the front contact under AM1.5G and one sun (1000 W/m^2). The material parameters for each layer were incorporated and varied accordingly as per the requirement of the analysis. In this study, the occurrence of PID is investigated for TFPV structures based on a variety of absorber layers with constant thickness. The effect of PID progression inside the simulated TFPV structures is qualitatively, as well quantitatively, analyzed in terms of their output performance metrics with appropriate justifications.

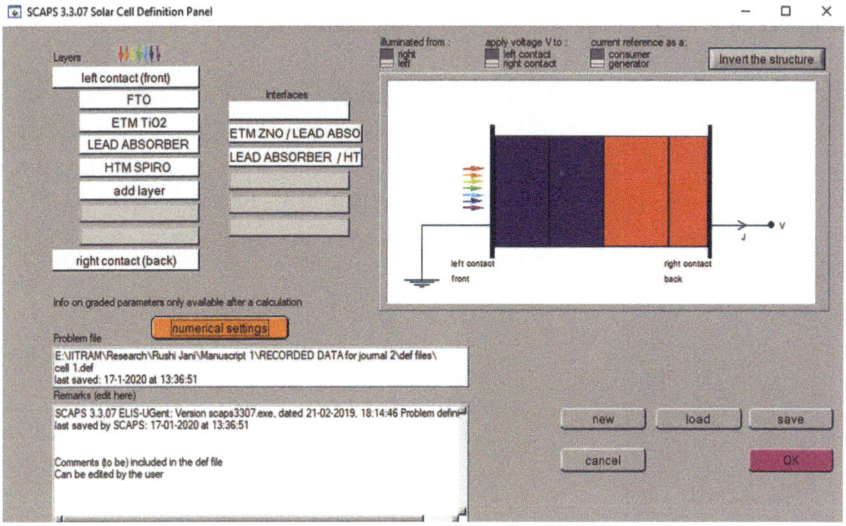

Figure 6.2 Overview of perovskite TFPV model in SCAPS.

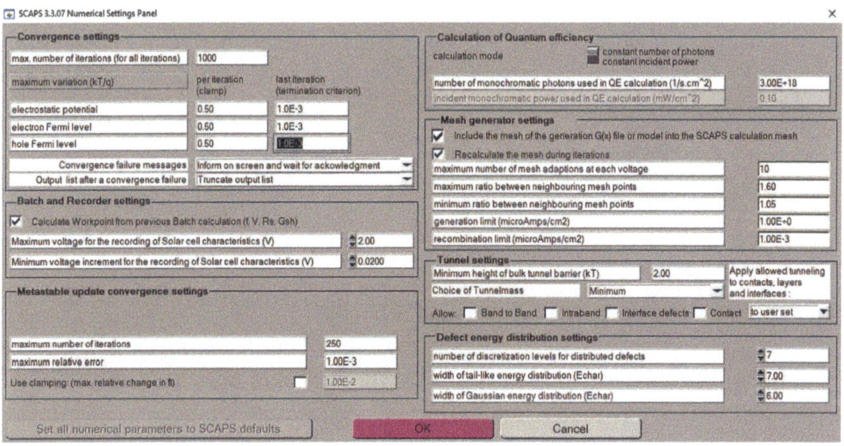

Figure 6.3 Numerical settings panel in SCAPS.

The complete simulation methodology adopted for the study is explained as follows. The procedure starts with the setting of the working point in SCAPS-1D. Further, the simulation parameters are defined and varied as per the requirement of the study. To study the PID effect in TFPVs, the defect densities have been accordingly varied inside the cells. The reason behind incorporating such a variation originates from the fact that the PID phenomena inside the cells create defect levels at several locations inside the cell.

However, defect states vary for different types of PV cells owing to the variations in the crystal properties generated due to variations in the deposition methods. To study the effect of PID at the cell level, the concentration of defect states is varied inside the bulk of the absorber layer, causing PID losses. The flowchart of the adopted simulation methodology is depicted in Fig. 6.4.

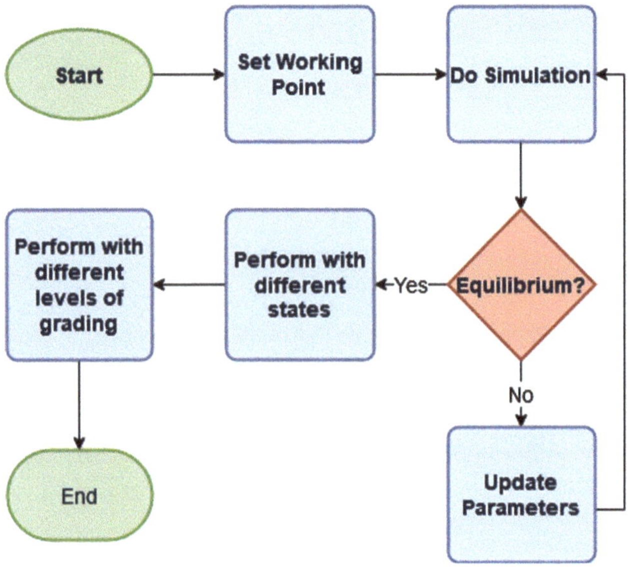

Figure 6.4 Flowchart of the simulation methodology.

The working point is initially set for the layer parameters, temperature, series resistance, shunt resistance, number of convergence points, and shunt resistances. After inputting the basic settings, the simulation is run until the equilibrium is not reached, which depends on the convergence of the model. In SCAPS-1D, the convergence failure is generally encountered and can be easily overcome through variations of the numerical settings and by checking the layer parameters incorporated into the model. The iterations are repeatedly done by changing the number of convergence points until the problem of nonconvergence is resolved.

Alternately, if nonconvergence is troubleshot through a correction in layer parameters, the simulation is repeated with the corrected parameters. In this study, the simulation results are described based on variable conditions of defect density, band-gap grading, etc., to analyze the effect of PID in different types of cells and their layer properties. Particularly, the variation of defect density is incorporated at different locations inside the simulated cell structures to mimic the scenario similar to that created due to PID progression. The adopted TFPV model is (Ni/Al)/MgF_2/ZnO:B/i-ZnO/CdS/OVC/Absorber/Mo/substrate as shown in Fig. 6.5 (Song et al. 2004) [11].

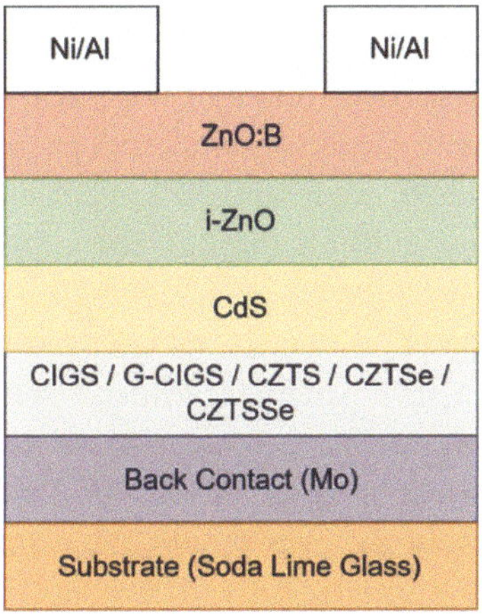

Figure 6.5 Cell structure of simulated thin-film solar cells with different absorber layers.

In a TFPV model, the most vital parts are the absorber layer, buffer layer, and an intermediate thin layer known as the ordered vacancy compound (OVC), which originated from the interface states residing in between the buffer absorber layer. Incorporating an OVC layer helps reduce the recombination rate because it shifts the p–n junction from the interface region, facilitating the high rate of recombination [12].

Generally, TFPVs use CIGS as an absorber layer and ZnO:Al as a Transparent Conducting Oxide (TCO) layer. However, boron-doped ZnO is used as a TCO because of comparatively lower absorption losses, leading to the improvement of the cells' quantum efficiency in and around the IR region. Also, TCO is covered with an anti-reflection coating (MgF_2) to improve the photon absorption inside the absorber layer. This happens because more photons will penetrate inside the cell structure due to the reduced extent of reflection from a cell structure's nonuniform top surface. As a result, the performance, reproducibility, and reliability of TFPV cells increases. Various absorber layer materials are considered to highlight the PID effect's dependence on absorber-layer properties in the present study.

Note that the thickness of all the used absorber layers is considered as 1 μm. The various absorber layers considered in the study are copper indium gallium selenide (CIGS), graded-CIGS (G-CIGS), copper zinc tin sulphide (CZTS), copper zinc tin selenide (CZTSe), and copper zinc tin sulphide selenide (CZTSSe).

The material parameters of CIGS and G-CIGS layers are taken from Song et al. (2004) [11]. Also, the parameters of CZTS, CZTSe, and CZTSSe layers are taken from Simya et al. (2015) [12]. Another assumption worth mentioning is that, except for the absorber layer, the rest of the layers are considered to be the same throughout the study.

Tables 6.1 and 6.2 show the list of simulation parameters adopted during the study, where (a) and (d) represent shallow acceptor and donor defect density. The sunlight transmission through the cells is kept at the default (99%) for each simulated TFPV structure. The defect states introduced inside each structure are used to model the progression of PID in TFPVs. The variation in performance parameter values is studied carefully for each simulated TFPVs by introducing the defect states at the bulk and interface locations inside the cells. It is observed that the incorporation of defects at different locations inside the cells would affect the efficiency almost in the same way. Still, in the interface, the defect level and location are case-sensitive [13].

Table 6.1 List of simulation parameters of TFSC absorber layers.

	CIGS	CZTS	CZTSe	CZTSSe
W (nm)	1000	1000	1000	1000
E_g (eV)	1.15	1.5	0.95	1.3
X (eV)	4.5	4.36	4.35	4.1
ϵ/ϵ_0	13.6	13.6	13.6	13.6
N_c $(1/cm^3)$	2.2×10^{18}	2.2×10^{18}	2.2×10^{18}	2.2×10^{18}
N_v $(1/cm^3)$	1.5×10^{19}	1.8×10^{19}	1.8×10^{19}	1.8×10^{19}
v_e (cm/s)	3.9×10^7	10^7	10^7	10^7
v_h (cm/s)	1.4×10^7	10^7	10^7	10^7
μ_e (cm^2/Vs)	100	100	100	100
μ_h (cm^2/Vs)	12.5	25	25	25
$Doping(1/cm^3)$	$10^{15}(a)$	$10^{18}(a)$	$5 \times 10^{16}(a)$	$10^{15}(a)$

Generally, the allocation of defect states inside the semiconductor causes a reduction in the generation of electron-hole pairs owing to the creation of deep (far away from conduction and valence band edges) energy levels

Table 6.2 List of simulation parameters for other TFSC layers.

	CdS	i-ZnO	ZnO:B
W (nm)	50	120	400
E_g (eV)	2.4	3.3	3.3
X (eV)	4.2	4.4	4.55
ϵ/ϵ_0	10	9	9
N_c (1/cm^3)	2.2×10^{18}	2.2×10^{18}	3×10^{18}
N_v (1/cm^3)	1.8×10^{18}	1.8×10^{19}	1.8×10^{19}
v_e (cm/s)	10^7	10^7	2.4×10^7
v_h (cm/s)	10^7	10^7	1.3×10^7
μ_e (cm^2/Vs)	100	100	100
μ_h (cm^2/Vs)	25	25	31
$Doping$(1/cm^3)	$10^{17}(d)$	$10^{18}(a)(d)$	$10^{20}(d)$

inside the forbidden gap. These energy levels act like trap states and facilitate the recombination of charge carriers, thus inducing the localized shunts (reduction in shunt resistance). PID progression is analogous to creating the localized shunt paths at different locations inside TFPV cells [5]. When produced inside the absorber layer, the defect states, buffer layer, or absorber/buffer interface degrades the performance of TFPVs the most. The effect of PID on TFPV performance begins to show when PID reaches the aforementioned locations inside the cell [2]. Different TFPV cells' performance is analyzed qualitatively and quantitatively using the SCAPS-1D tool for variable defect density values in the range 10^{14}–10^{19} cm^{-3}.

This chapter also discusses the PID phenomenon's impact on band-gap graded TFPV cells. The level of grading enhances the performance of TFPVs explicitly. The effect of defect density on G-CIGS cell-performance metrics is also simulated to mimic the effect of PID progression in these cells with the intuition of achieving lesser performance degradation compared to nongraded CIGS cells. The G-CIGS layer is incorporated in the following manner. The CIGS absorber is divided into two layers: The first layer with a uniform band gap is defined close to the pn junction; the second linearly graded band-gap layer (comparatively thinner) is defined near to back contact. The band gap of the first layer is kept constant throughout at 1.2 eV. In comparison, the linearly graded region gradually increases from 1.2 eV to 1.4 eV to locate the perfectly synchronized spot. Furthermore, the thickness of the linearly graded region band section (d_{grd}) is varied from 0.1 to 0.5 μm, while the thickness of the uniform band-gap region is kept constant at 1 μm. The variation in the position of the conduction band edge varies with the band gap, which changes due to the grading of gallium content in CIGS thin-film (compositional grading). Graded composition of the TFPV absorber layer results in the grading of its optoelectronic parameters, such as the energy band gap Eg(x), electron affinity χ(x), optical absorption $\alpha(x, \lambda)$, effective density of states $N_C(x)/N_V(x)$, doping density $N_D(x)/N_A(x)$, carrier mobility $\mu_n(x)/\mu_p(x)$, and recombination properties $N_t(x)$, among others.

6.3 Simulation-based analysis of PID effects in TFPV cells

The next phase post creation of various types of TFPV models in the SCAPS-1D tool. The defect density is varied inside the absorber layer of each simulated cell. It would create a performance degrading effect similar to that induced by the PID phenomenon. In this context, firstly, the aforesaid is done and analyzed for nongraded and graded type CIGS TFPV models. Secondly, a similar scheme was applied over CZTS-, CZTSe-, and CZTSSe-based TFPV models. The variations in the output performance metrics of each simulated cell are studied in detail. It is worth mentioning here that compositional grading of the absorber layer in CZTS-based TFPVs is recommended to increase their conversion efficiency. This can be achieved by selective introduction of selenium, decreasing sulfur content's percentage inside the crystal structure. The resulting absorber-layer material is CZTSSe, which has superior optoelectronic parameters that prove to be more effective in enhancing the conversion efficiency TFPVs. Afterward, a similar PID mechanism introduction is also done for the remaining CZTS-, CZTSe-, and CZTSSe-based TFPV models. Note that the impact of defect states in all the layers except the absorber layer is negligible. The defect density is only increased inside the absorber layer of each model. In this way, PID progression is studied relative to absorber layer materials, i.e., CIGS, G-CIGS, CZTS, CZTSe, and CZTSSe.

The relative assessment of PID progression in all the TFPV cells is based on energy-band diagrams and quantum efficiency curves. The variation of defect density is supposed to have a profound impact on these. Actually, defect states create trap sites inside the forbidden energy gap (band gap), which act as recombination centers and cause reduction of charge-carrier collection at the electrodes. With the advent of PID, some drastic changes in the energy-band diagrams (Figs. 6.4, 6.10, 6.12, 6.14, and 6.16) and quantum efficiency curves (Figs. 6.3, 6.9, 6.11, 6.13, and 6.15) for all cells are observed and analyzed meticulously. The analysis highlights the utility of performance metrics, quantum efficiency curves, and energy-band diagrams to understand and predict the magnitude of PID degradation in TFPV cells. A detailed analysis is provided in the following subsections.

6.3.1 Effect of PID in nongraded CIGS cells

The effect of PID onset is first analyzed for a nongraded CIGS TFPV cell. It is observed that variations in the defect density affects the performance of cells significantly. It is well known that PID progression generates defect states similar to those created by voids, doping, surfaces, or interfaces inside semiconductor crystals. The result of PID is always evident in terms of reduction in the values of V_{OC} and J_{SC}. Such decreased performance parameters of TFPVs are practically significant. The presence of defect states inside the absorber or buffer layer's band gap causes discrete variation in V_{OC} values for different types of TFPV cells. Fig. 6.6 shows a remarkable change in the value of band-gap energy with the advent of PID.

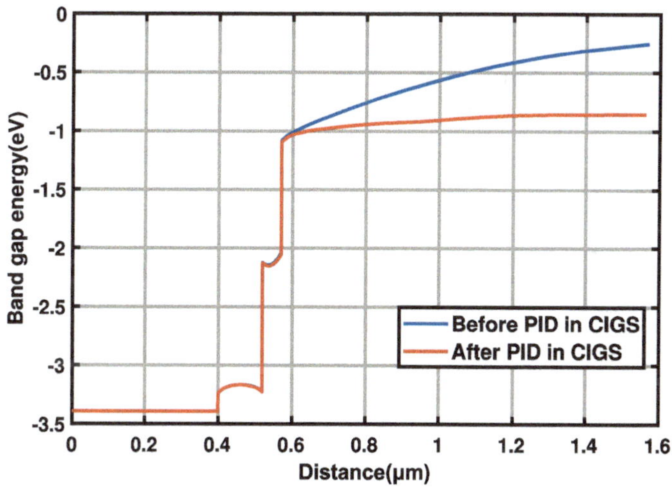

Figure 6.6 Energy band diagram of nongraded CIGS solar cell.

The initial value of band-gap energy of −3.39 eV is modified to −0.85 eV attributed to PID. Without PID, the final value of band-gap energy is found to be −0.25 eV. Also, the value of quantum efficiency for nongraded CIGS cells without PID increased from 3.41% to 80.94%. In contrast, the onset of PID causes a shift in its value from 0.43% to 0.08% as shown in

Fig. 6.7. To mimic the scenario of the variable magnitude of PID progression inside TFPV cells, the defect density value is varied from 10^{14} cm^{-3} to 10^{19} cm^{-3}. The reason behind the selection of such a range is that, with the advent of PID, the performance degradation is minimal initially, but it grows severely after some time. As mentioned in Chapter 3, PID produces degradation in V_{OC}. Therefore, the effect of the variable magnitude of PID is measured in terms of V_{OC} values. Fig. 6.8 shows that V_{OC} decreases from 0.56 V to 0.11 V.

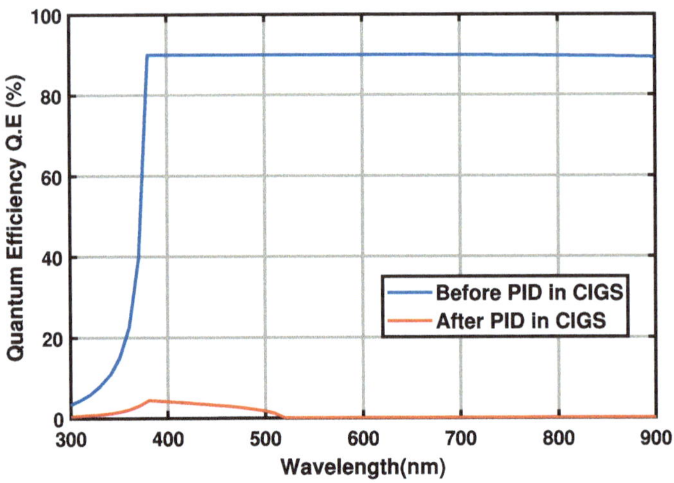

Figure 6.7 Plot of QE-λ of nongraded CIGS solar cells.

Also, a reduction in the value of J_{SC} is expected, which is also evident from simulation results. J_{SC} is observed to decrease from 37.4 mA/cm^2 to 0.12 mA/cm^2 (Fig. 6.9). Similarly, in Fig. 6.10, the fill-factor value also shows a reduction from 78.66% to 41.03%.

The most important reduction is visible in the efficiency value from 16.34% to 0.01% (Fig. 6.11). From these observations, we conclude that performance-parameter values of nongraded CIGS cells become almost negligible when the PID effect is severe.

Figs. 6.8–6.11 also signify changes in the values of performance metrics of simulated CZTS, CZTSe and CZTSSe TFPV cells. The conversion efficiency of CZTS TFPV declines from 10.16% to 0.52% as the magnitude of PID increases to become more severe (Fig. 6.11).

J_{SC} reduces from 16.11 mA/cm^2 to 2.52 mA/cm^2 (Fig. 6.9), but, contrarily, the fill-factor improved from 11.81% to 60% as seen in Fig. 6.10,

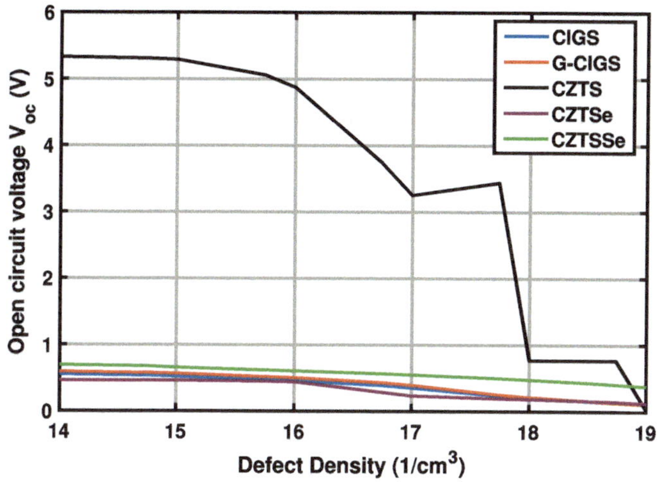

Figure 6.8 Comparison of V_{OC} vs. defect density for various simulated TFSCs.

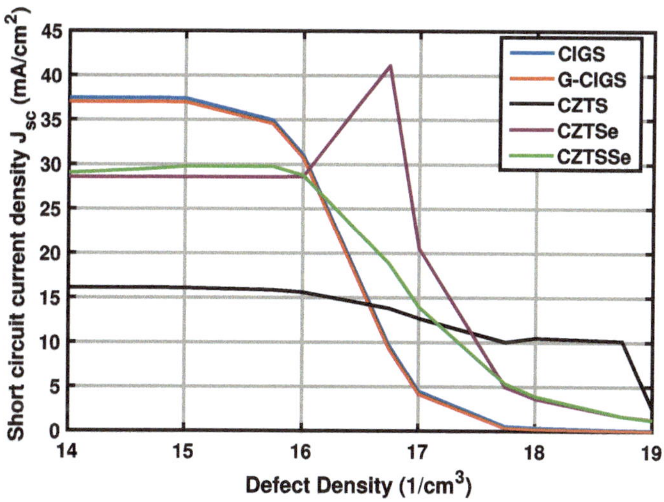

Figure 6.9 Comparison of J_{SC} vs. defect density for various simulated TFSCs.

and V_{OC} suffers reduction from 5.33 V to 0.03 V (Fig. 6.8). In a CZTSe cell, the V_{OC} reduces from 0.47 V to 0.13 V (Fig. 6.8), J_{SC} experiences lowering in its magnitude from 28.59 mA/cm^2 to 1.31 mA/cm^2 (Fig. 6.9), efficiency shows a decrement from 10.28% to 0.09% (Fig. 6.11), and the fill-factor reduces from 77.07% to 48.68% (Fig. 6.10). Finally, the cell with

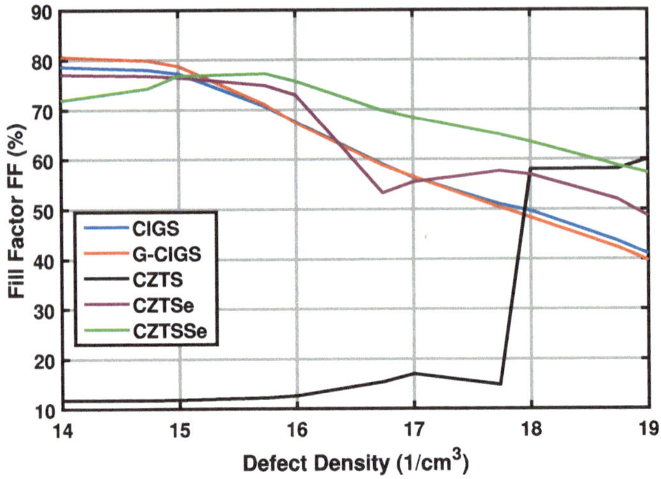

Figure 6.10 Comparison of FF vs. defect density for various simulated TFSCs.

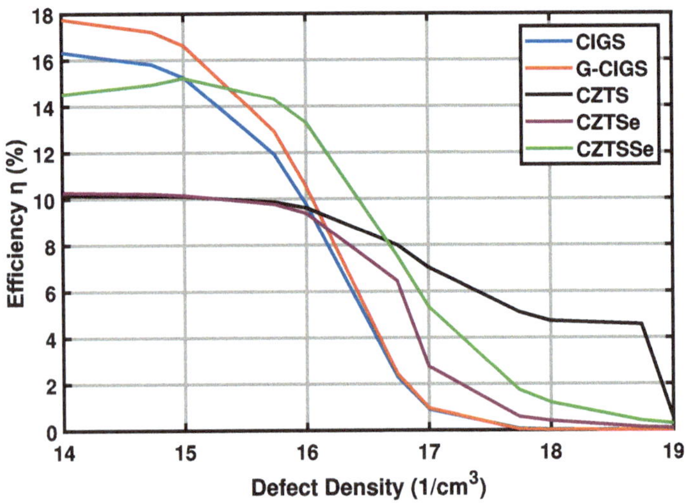

Figure 6.11 Comparison of η vs. defect density for various simulated TFSCs.

a completely back-graded absorber layer, i.e., a CZTSSe TFPV cell, is simulated to find the impact of PID on its output parameters.

It is noteworthy that the conversion efficiency of a CZTSSe cell is much superior to CZTS and CZTSe cells, which is attributed to the effect of compositional grading achieved experimentally through the proportionate replacement of sulphur with selenium. Further, the effect of increasing defect density (PID effect) is also studied for a CZTSSe cell. In this case, too, the impact of PID onset is quite devastating in terms of the reduction in the values of all the performance metrics. The value of VOC deteriorates from 0.69 V to 0.39 V (Fig. 6.8), JSC reduces from 29.05 mA/cm^2 to 1.27 mA/cm^2 (Fig. 6.9), the fill-factor drops from 71.93% to 57.27% (Fig. 6.10), and the conversion efficiency also degrades from 14.53% to 0.28% (Fig. 6.11).

The aforementioned cluster of performance-parameter values strongly suggests about the negative impact of PID onset on the performance of all the simulated TFPV cells. Therefore, irrespective of the fact that what absorber-layer material has been used to fabricate the TFPV cells, the advent of PID phenomenon tends to severely degrade the performance. However, the relative impact of PID is observed to be dependent on the properties of the absorber layer, and it could be one of the counter strategies to reduce PID in cells but not the most effective one for sure. Further, it is to be noted that the maximum conversion efficiency for nongraded CIGS cell is 16.53%, whereas for a G-CIGS cell it increases to 17.96%. With the onset of PID, the efficiency of a G-CIGS cell decreases from 17.75% to 0.001% as shown in Fig. 6.11. Also, V_{OC} for G-CIGS is reduced from 0.59 V to 0.11 V (Fig. 6.8), J_{SC} is also reduced from 37.02 mA/cm^2 to 0.06 mA/cm^2 (Fig. 6.9), and the fill-factor drops from 80.60% to 39.68% (Fig. 6.10).

6.3.2 Effect of PID in graded CIGS cells

The potential of G-CIGS TFPV cells in countering the onset of PID progression is evaluated. For this purpose, the absorber layer with a nonuniform band gap is incorporated to create a G-CIGS cell, and its efficacy to counter the advent of PID is investigated. Experimentally, this is achieved by introducing the compositional variations in the gallium content.

In this study, a highly efficient G-CIGS cell is studied with back grading at thickness ($d_g = 0.1$ μm). Note that the incorporated back contact grading is linear in nature. Fig. 6.12 shows the change in band-gap value of G-CIGS cells from −3.39 eV to −0.26 eV (without PID) and −0.85 eV (with PID). The increment in quantum efficiency from 3.41% to 80.93% (without PID) is modified as 0.15% to 0.08% (with PID) as shown in Fig. 6.13.

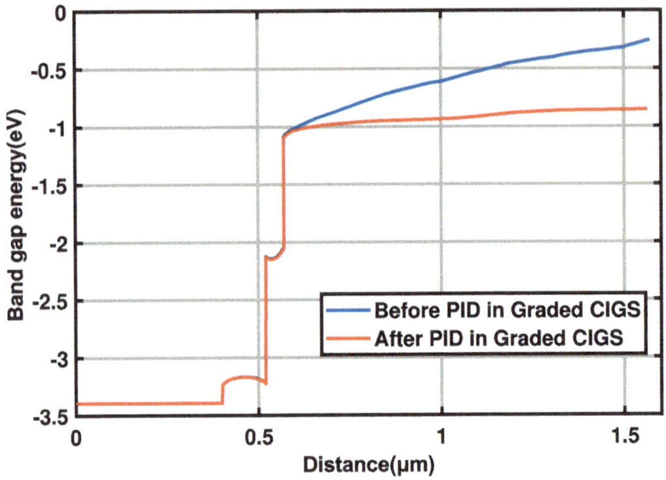

Figure 6.12 Energy-band diagram of a graded CIGS solar cell.

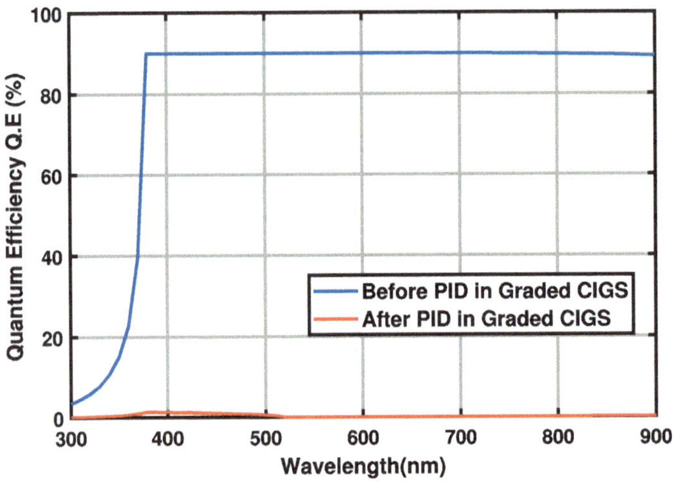

Figure 6.13 Plot of QE-λ of a graded CIGS solar cell.

6.3.3 Effect of PID in CZTS, CZTSe and CZTSSe cells

The impact of PID onset is also evident in simulated CZTS, CZTSe and CZTSSe TFPV cells as quantitatively demonstrated in Figs. 6.14, 6.16 and 6.18 in terms of energy band-gap values.

The band-gap energy values for CZTS, CZTSe and CZTSSe cells are changed from −3.39 eV to −0.67 eV, −0.75 eV and −0.85 eV respectively.

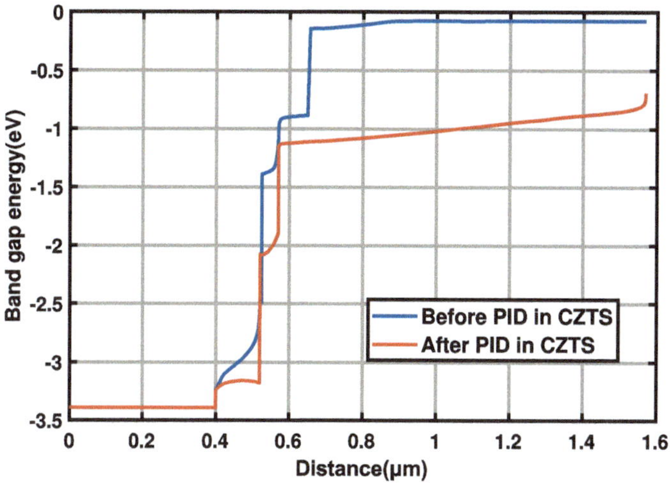

Figure 6.14 Energy-band diagram of CZTS solar cell.

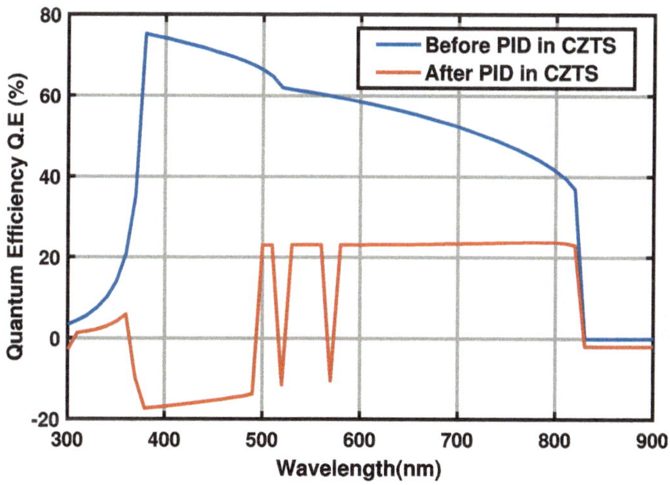

Figure 6.15 Plot of QE-λ of a CZTS solar cell.

The quantum efficiency for CZTS is reduced drastically in most of the visible region with the onset of PID as shown in Fig. 6.15.

Also, the PID variation is transient in nature after onset. A similar kind of effect on quantum efficiency distribution is also observed in a CZTSe cell. The distribution of quantum efficiency ranges from 3.33% to 50.41% (without PID) and 5.16% to 1.09% (with PID) (Fig. 6.17).

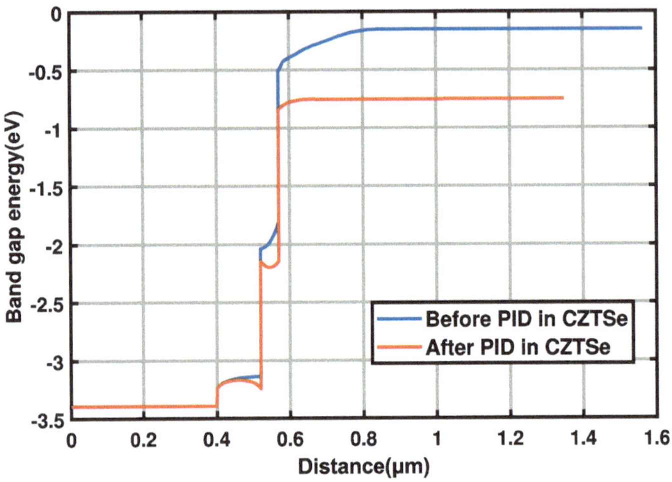

Figure 6.16 Energy-band diagram of a CZTSe solar cell.

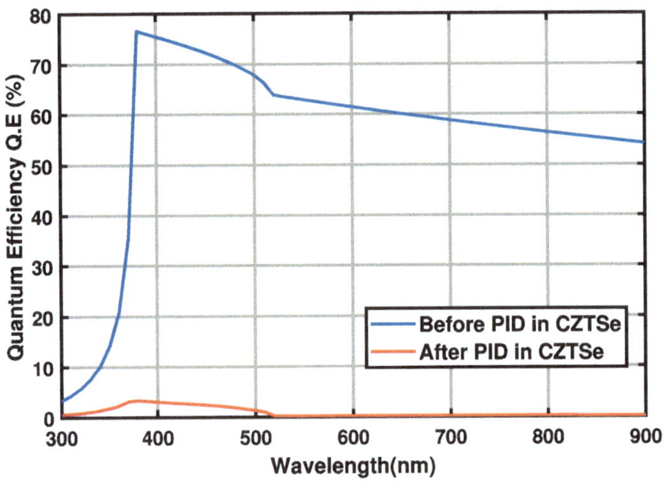

Figure 6.17 Plot of QE-λ of a CZTSe solar cell.

In the case of CZTSSe cells, the quantum-efficiency variation is initially between 3.4% to 70.71% (without PID), but 0.34% to 3.34% in a PID-affected cell (Fig. 6.19).

The more precise values of the performance parameters are tabulated in Tables 6.3 to 6.6.

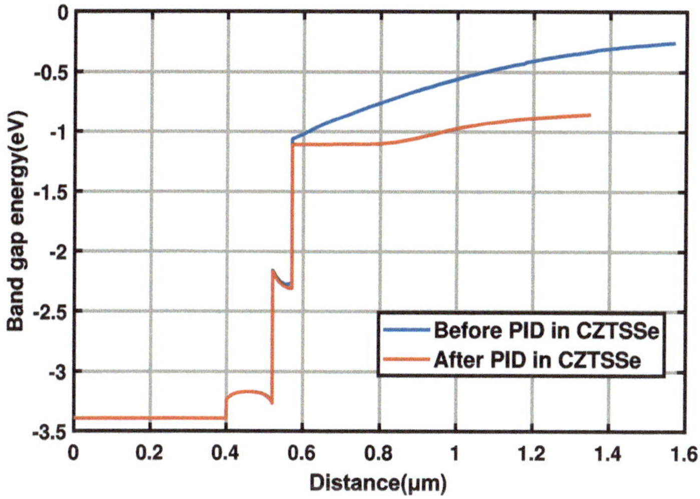

Figure 6.18 Energy-band diagram of CZTSSe solar cell.

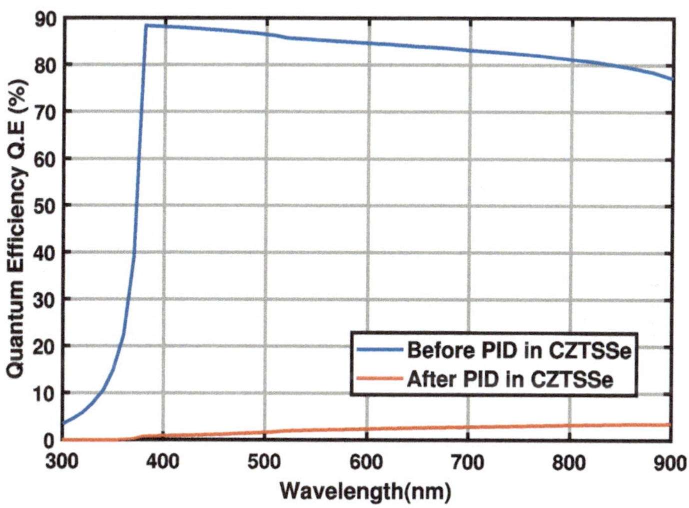

Figure 6.19 Plot of QE-λ of a CZTSSe solar cell.

The tabulated values in these cases suggest that reduction in the conversion efficiency and open-circuit voltage are the two most affected performance metrics in the advent of PID in cells.

Table 6.3 List of output open-circuit voltage parameters for simulated TFSCs.

Defect density	Open circuit voltage V_{oc} (V)				
	CIGS	G-CIGS	CZTS	CZTSe	CZTSSe
1×10^{14}	0.5552	0.5948	5.3301	0.4663	0.6953
5.5×10^{14}	0.5412	0.5826	5.3110	0.4655	0.6813
1×10^{15}	0.5273	0.5702	5.2901	0.4647	0.6653
5.5×10^{15}	0.4817	0.5256	5.0602	0.4568	0.6226
1×10^{16}	0.4652	0.5089	4.8700	0.4486	0.6105
5.5×10^{16}	0.3965	0.4382	3.7403	0.2940	0.5716
1×10^{17}	0.3586	0.3981	3.2500	0.2410	0.5551
5.5×10^{17}	0.2424	0.2578	3.4401	0.2027	0.5032
1×10^{18}	0.2128	0.2224	0.7701	0.1919	0.4831
5.5×10^{18}	0.1381	0.1401	0.7703	0.1511	0.4133
1×10^{19}	0.1144	0.1147	0.0293	0.1301	0.385

Table 6.4 List of output short-circuit current-density parameters for simulated TFSCs.

Defect density	Short circuit current density J_{sc} (mA/cm^2)				
	CIGS	G-CIGS	CZTS	CZTSe	CZTSSe
1×10^{14}	37.45	37.02	16.11	28.59	29.05
5.5×10^{14}	37.44	37.02	16.09	28.58	29.51
1×10^{15}	37.43	36.98	16.06	28.58	29.77
5.5×10^{15}	34.97	34.57	15.83	28.58	29.77
1×10^{16}	31.07	30.68	15.60	28.62	28.77
5.5×10^{16}	9.78	9.37	13.84	41.15	18.87
1×10^{17}	4.49	4.17	12.67	20.55	13.98
5.5×10^{17}	0.598	0.353	10.00	4.95	5.3
1×10^{18}	0.411	0.216	10.46	3.65	3.9
5.5×10^{18}	0.165	0.079	10.15	1.7	1.69
1×10^{19}	0.122	0.057	2.52	1.31	1.27

Although the adverse effect on efficiency seems to be quite obvious, the percentage of reduction in the values of open-circuit voltages is visualized in Fig. 6.17 for all the TFPV cells considered in this study.

Another interesting observation is that the maximum degradation of 99.45% is found in a CZTS cell, while the minimum percentage reduction of 44.62% due to PID progression is found in a CZTSSe cell (see Fig. 6.20).

Table 6.5 List of output efficiency parameters for simulated TFSCs.

Defect density	**Efficiency η (%)**				
	CIGS	**G-CIGS**	**CZTS**	**CZTSe**	**CZTSSe**
1×10^{14}	16.34	17.75	10.16	10.28	14.53
5.5×10^{14}	15.81	17.21	10.13	10.22	14.94
1×10^{15}	15.24	16.61	10.11	10.16	15.22
5.5×10^{15}	11.91	12.91	9.87	9.77	14.31
1×10^{16}	9.76	10.52	9.62	9.37	13.3
5.5×10^{16}	2.29	2.42	8	6.45	7.53
1×10^{17}	0.91	0.94	7.02	2.75	5.31
5.5×10^{17}	0.07	0.05	5.11	0.58	1.74
1×10^{18}	0.04	0.02	4.72	0.4	1.2
5.5×10^{18}	0.01	0.001	4.57	0.13	0.41
1×10^{19}	0.01	0.001	0.52	0.09	0.28

Table 6.6 List of output fill-factor parameters for simulated TFSCs.

Defect density	**Fill factor FF (%)**				
	CIGS	**G-CIGS**	**CZTS**	**CZTSe**	**CZTSSe**
1×10^{14}	78.66	80.6	11.81	77.07	71.93
5.5×10^{14}	77.99	79.79	11.85	76.78	74.32
1×10^{15}	77.22	78.75	11.89	76.47	76.81
5.5×10^{15}	70.68	71.04	12.30	74.86	77.23
1×10^{16}	67.53	67.35	12.68	72.93	75.72
5.5×10^{16}	59.06	58.83	15.43	53.27	69.78
1×10^{17}	56.39	56.5	17.02	55.48	68.4
5.5×10^{17}	51.05	50.41	14.83	57.67	64.97
1×10^{18}	49.7	48.3	58.05	56.95	63.48
5.5×10^{18}	43.66	42.36	58.08	52.02	58.76
1×10^{19}	41.03	39.68	60	48.68	57.27

In this chapter, we presented the results of a computational analysis of the impact of PID phenomenon on various popular TFPV cells, which, if done through fabrication of cells or modules, would result in the wastage of both time and expensive fabrication resources. The novelty of the presented analysis lies in the fact that the progression of defect states was done analogously to the progression of PID effects in TFPV cells based on different absorber-layer materials. Additionally, the effects of PID were also presented in terms of band-gap energy distribution and quantum-efficiency distribu-

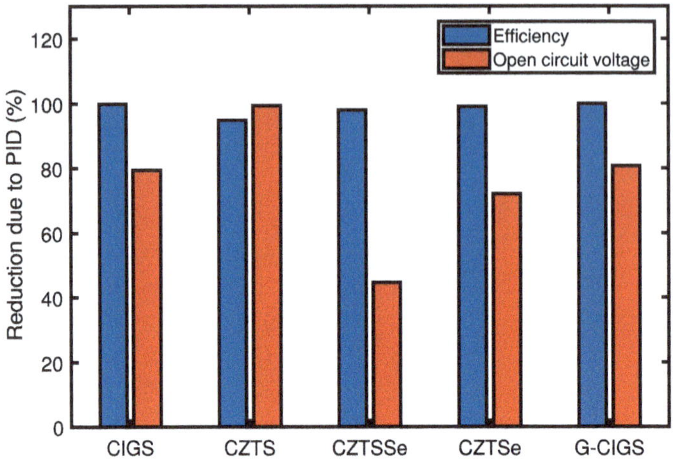

Figure 6.20 PID effect on various thin-film solar cells.

tion. The analysis is mainly helpful in understanding the PID phenomenon at the cell level.

The progression of PID was observed to be critical and will cause more damage if it persists longer. The most affected performance due to PID was seen in G-CIGS cell, in which the conversion efficiency was reduced by 99.99%, and the minimum efficiency reduction of 94.88% was found in CZTS cell. The changes in V_{OC}, which is the most prominent result of PID, was also monitored. The maximum decrease of Voc was found in a CZTS cell (99.45%), whereas it was minimum for a CZTSSe cell (44.62%). The comparison of results suggests that the order of vulnerability levels to PID occurrence in various TFPV was in this order:

$$\text{CZTS} > \text{G-CIGS} > \text{CIGS} > \text{CZTSe} > \text{CZTSSe}$$

Thus, it can be concluded that a CZTSSe TFPV is least prone to PID, whereas a CZTS TFPV is the most affected. One can further extend this analysis for TFPVs based on other advanced absorber-layer materials which would be helpful in developing more reliable TFPV technology. Also, the investigation can be further extended to compare the simulation results with real-time data obtained experimentally using several sophisticated tools and equipment.

As solar power becomes more extensively used worldwide, it becomes clear that the primary problem of solar PV technology and effective solar-power systems is a power loss due to the unavoidable PID effect. Numerous factors cause the PID effect in the solar-power system. An anti-PID device is a critical component of any commercial solar-power system or plant that consists of solar PV panels and inverters. Installation of a professional anti-PID solution ensures maximum efficiency and the highest yield of solar power, as well as longer system lifetimes with reduced maintenance or a need for panel replacement [14].

Bibliography

[1] N.C. Dong, M.A. Islam, Y. Ishikawa, Y. Uraoka, The influence of sodium ions decorated micro-cracks on the evolution of potential induced degradation in p-type crystalline silicon solar cells, Solar Energy 174 (2018) 1–6, https://doi.org/10.1016/j.solener.2018.08.082.

[2] S. Yamaguchi, K. Ohdaira, Degradation behavior of crystalline silicon solar cells in a cell-level potential-induced degradation test, Solar Energy 155 (2017) 739–744, https://doi.org/10.1016/j.solener.2017.07.009.

[3] S. Yamaguchi, S. Jonai, K. Hara, H. Komaki, Y. Shimizu-Kamikawa, H. Shibata, S. Niki, Y. Kawakami, A. Masuda, Potential-induced degradation of Cu(In, Ga)Se_2 photovoltaic modules, Japanese Journal of Applied Physics 54 (8S1) (2015) 08KC13, https://doi.org/10.7567/jjap.54.08kc13.

[4] S. Steingrube, O. Breitenstein, K. Ramspeck, S. Glunz, A. Schenk, P.P. Altermatt, Explanation of commonly observed shunt currents in c-Si solar cells by means of recombination statistics beyond the Shockley-Read-Hall approximation, Journal of Applied Physics 110 (1) (2011) 014515, https://doi.org/10.1063/1.3607310.

[5] V. Naumann, D. Lausch, S. Großer, M. Werner, S. Swatek, C. Hagendorf, J. Bagdahn, Microstructural analysis of crystal defects leading to potential-induced degradation (PID) of Si solar cells, Energy Procedia 33 (2013) 76–83, https://doi.org/10.1016/j.egypro.2013.05.042.

[6] P.M. Salomé, V. Fjällström, P. Szaniawski, J.P. Leitão, A. Hultqvist, P.A. Fernandes, J.P. Teixeira, B.P. Falcão, U. Zimmermann, A.F. da Cunha, M. Edoff, A comparison between thin film solar cells made from co-evaporated $CuIn_{1-x}Ga_xSe_2$ using a one-stage process versus a three-stage process, Progress in Photovoltaics: Research and Applications 23 (4) (2014) 470–478, https://doi.org/10.1002/pip.2453.

[7] C.-S. Jiang, C. Xiao, H. Moutinho, S. Johnston, M. Al-Jassim, X. Yang, Y. Chen, J. Ye, Imaging charge carriers in potential-induced degradation defects of c-Si solar cells by scanning capacitance microscopy, Solar Energy 162 (2018) 330–335, https://doi.org/10.1016/j.solener.2017.12.025.

[8] S. Mishra, K. Bhargava, D. Deb, Numerical simulation of potential induced degradation (PID) in different thin-film solar cells using SCAPS-1d, Solar Energy 188 (2019) 353–360, https://doi.org/10.1016/j.solener.2019.05.077.

[9] M. Burgelman, P. Nollet, S. Degrave, Modelling polycrystalline semiconductor solar cells, Thin Solid Films 361–362 (2000) 527–532, https://doi.org/10.1016/s0040-6090(99)00825-1.

[10] R. Jani, K. Bhargava, Comparative investigation into effects of the interplay between absorber layer crystallinity and interfacial defect states on the performance of lead-based and tin-based perovskite solar cells, Semiconductor Science and Technology 35 (10) (2020) 105007, https://doi.org/10.1088/1361-6641/aba229.

[11] J. Song, S.S. Li, C. Huang, O. Crisalle, T. Anderson, Device modeling and simulation of the performance of $Cu(In_{1-x},Ga_x)Se_2$ solar cells, Solid-State Electronics 48 (1) (2004) 73–79, https://doi.org/10.1016/s0038-1101(03)00289-2.

[12] O. Simya, A. Mahaboobbatcha, K. Balachander, A comparative study on the performance of Kesterite based thin film solar cells using SCAPS simulation program, Superlattices and Microstructures 82 (2015) 248–261, https://doi.org/10.1016/j.spmi.2015.02.020.
[13] F. Wan, F. Tang, H. Xue, W. Lu, Y. Feng, Z. Rui, Effects of defect states on the performance of $CuInGaSe_2$ solar cells, Journal of Semiconductors 35 (2) (2014) 024011, https://doi.org/10.1088/1674-4926/35/2/024011.
[14] D. Deb, N.L. Brahmbhatt, Review of yield increase of solar panels through soiling prevention, and a proposed water-free automated cleaning solution, Renewable & Sustainable Energy Reviews 82 (2018) 3306–3313, https://doi.org/10.1016/j.rser.2017.10.014.

CHAPTER 7

PID for multicrystalline soiled panels: a forecasting-based approach

Due to the inconsistency and uncertainty concerning photovoltaic (PV) yields, a precise forecast of PV-output parameters is tricky. Framing an optimal forecasting approach to predict the reduction in the panel currents will help adopt appropriate Potential-Induced Degradation (PID)-recovery techniques. To anticipate a PV panel's current by distinctive forecasting models, specific parameters such as temperature, humidity, and soiling levels are definite PID indicators. Mainly, we investigate the PID's effect by introducing leakage currents due to the combination of atmospheric attributes and soiling levels. We are undertaking experiments at a solar plant located in Nimbahera, Rajasthan, India. This paper employs forecasting techniques consisting of the Persistence Learning Algorithm (PLA), Auto-Regressive Integrated Moving Average (ARIMA), Single Exponential Smoothing (SES), Artificial Neural Network (ANN), Support Vector Regression (SVR), Random Forests (RF), and Gradient boosted machine (GBM) for early prediction of panel currents. We perform the forecasting and numerical calculations using MATLAB® software and Statistical Package for the Social Sciences (SPSS) tool.

7.1 Introduction

With a significant leap in the establishment of sustainable power sources, fossil-fuel usage is being reduced. A vast majority of ventures in many areas lean towards sustainable power sources like solar and wind because of lower carbon footprints, and these are managed using advanced control techniques [1]. With solar energy abundantly available, tapping power efficiently from solar panels is a critical task. Due to such high demand, the adaptibility of crystalline-type solar cells is advancing continuously. The solar modules are an excellent resource due to certain disadvantages. A well-known degradation phenomenon is Potential Induced Degradation (PID), occurring in solar cells due to the migration of sodium ions through

Degradation, Mitigation, and Forecasting Approaches in Thin Film Photovoltaics
https://doi.org/10.1016/B978-0-12-823483-9.00018-8

the potential gradient from the front SLG substrate inside the anti-reflective coating (ARC) of a PV module.

Numerous factors affect the pace of PID in the crystalline-type solar cells, like humidity and temperature levels [2,3], illumination levels [4], recovery history [5,6], the voltage potential of the active cell circuit and system grounding mechanisms [7]. A significant amount of PID growth happens when the surface layer electrically conducts, leading to the formation of the ground electrode against the negative bias of cell [8,9]. The surface-layer soiling contributes to the harmful effect of PID on the solar panels. Soiling absorbs the humidity like a sponge, which leads to persistent surface groundings due to high leakage currents. The presence of soiling on module glass leads to diminished module surface resistance, further extending the potential across a grounded PV-module frame's glass surface. Various control strategies and cleaning methods are explored for electrical output reliability from the solar panels so far [10]. Also, at the cell level, the effect of PID is studied for various thin-film solar cells by introducing defect states at specific layers [11]. We extend this study to the PV-system level to estimate optimal PID detection and potential recovery techniques in the present work.

Electrical-power personnel should ensure a balance between the power generation and utilization at one time, but it is difficult to maintain the traditional and controllable energy production framework. Several nations are considering utilizing sustainable power sources in their power systems these days, creating challenging issues because the renewable sources (wind, solar, and so forth.) are unreliable. It is essential to anticipate the solar-output data adequately for a highly coordinated energy system. The biggest challenge for the new era is the successful integration of renewable-energy power sources. The power system's overall reliability depends on the system's capability to account for the unexpected variations and interruptions while ensuring quality and congruity of service to the clients. The combination of renewable energy sources with an existing power network escalates the grid system's complexity and the consistency of the production/consumption continuity due to their irregular nature [12,13]. The uncertainty in solar devices' power production includes other obstacles like voltage glitches, instability, power grade, and integration issues [14,15]. Due to such a gradual shift to renewable energy sources, the precise prediction of solar outputs becomes predominant for highly coordinated power framework [16]. In this work, we forecast the panel current, which helps in an early estimation of an adverse effect of PID for suitable recovery and consequent corrective steps.

Comparatively, lesser progress has taken place to determine the correlation between the influence of PID and soiling levels. The conductivity of the surface of the active cell circuit over the glass front affects the process of PID due to the grounded paths [17]. Due to the conductivity of soil deposited on the panel, the surface resistance of module glass is reduced [18]. The existence of surface soiling depends upon the ambient parameters and results in increased leakage currents in case of a lesser surface humidity magnitude. For PV modules packaged with a relatively low front-face bulk resistivity, the ionic leakage traversing to the module ends preferably for a variety of front-surface lateral resistances [19]. Utilization of more volume resistivity materials causes more uniform distribution of leakage currents for a PID-affected module [20].

The leakage currents get generated on the glass surface due to increased potential between the frame and the ground. The PID rate may have an immediate direct correlation with leakage current for a given PV module. The leakage currents induced are more likely to lead to PID losses. Earlier, researchers explored the effect of temperature, relative humidity, bias voltage, and contact to the ground on the leakage current's value [21]. There exists an Arrhenius relationship between temperature and leakage current for an activation energy magnitude of 75 $kJ\,mol^{-1}$. Also, an ohmic correlation exists between leakage current and bias voltage. The leakage current is lesser for low surface humidity and higher with a contacting metal foil at the surface. As discussed earlier, the leakage current in between the cells and frame faces ambient humidity. Due to the humidity, a thin layer develops on the surface glazing front which conducts electricity beyond a particular percolation threshold magnitude.

In this work, we forecast panel currents after conducting the soiling tests. Various forecasting models are used to anticipate future values. As the panel currents decrease due to leakage currents caused by PID, such investigations help assess the panel currents' future variations due to PID's impact. The panel currents encounter PID losses because of such induced leakage currents due to soiling. This chapter supplies a rigorous explanation of the ongoing and advanced forecasting plans for the panel currents in a PID-affected solar panel that prompts early estimation of PID and secures PID-recuperation measures in favor of healthy solar panels.

The expanded awareness of solar PV panels as a sustainable power source has raised the quantity of PV panel installations of late. Meanwhile, more accessibility of information and computational power have empowered machine-learning calculations for improved forecasts. Because predicting PV performance parameters is fundamental for the whole energy sector, forecasting techniques are essential to understand the effect of PID on output panel currents. Several machine learning and statistical methods can be incorporated for solar forecasting [22]. This examination compares various PV performance parameters and time-series forecasting models on a solar farm. We need to monitor the solar cells before incorporating them into the solar farm. Several factors affect the depth of PID in solar cells, mainly the ambient parameters such as temperature and relative humidity. The effects of ambient parameters need to be studied to obtain knowledge of PID dependence.

Earlier, machine learning techniques were explored and evaluated in combination with various features targeted to ameliorate the overall performance [23]. The time series forecasting methods are somewhat complicated due to the nonstationary energy time series. We study the general behavior of the PV panel current during the variation in the ambient parameters apart from the soiling-level changes. Apart from yielding the optimal forecasting method for panel current forecasting, the study will also help identify the core relationship between PID and soiling levels dependent on the wind speed and relative humidity. We also study other performance parameters for a better understanding of PID. The PID here results from soiling level, temperature, and relative humidity. Identifying the correlation between the studied parameters helps determine the actual effect of PID, followed by various forecasting methods for further analysis. The work mainly aims to find the optimal forecasting technique for estimating the critical nature of PID on the PV panel currents.

7.2 Experimental setup and methodology

Two panels of model number REC320PE 72 of $77.5 \times 39 \times 1.7$ in^3 dimensions, weighing 59.5 lbs and covering area of 1.95 m^2 from Renewable Energy Corporation (REC) are studied [24]. The nominal power P_{mpp} of 320 W with panel efficiency of 16.4% exists under standard test conditions STC (airmass AM 1.5, irradiance 1000 Wm^{-2}, 25°C cell temperature) for the panels. These panels consist of 72 multicrystalline cells, each with three strings of 24 cells along with bypass diodes. Although the operators claim the panel is PID-free, in practice, it doesn't mean that no degradation happened. Under the International Electrotechnical Commission (IEC) 62804 standard, modules are rated PID-free only if the power loss is less than 5% [25]. Hence, PID losses less than 5% remain unscrutinized and may increase substantially with time [26]. The study conducted during winter at a solar plant situated at Nimbahera, Rajasthan, India, containing two panels, as shown in Fig. 7.1, is subjected to soiling-related tests.

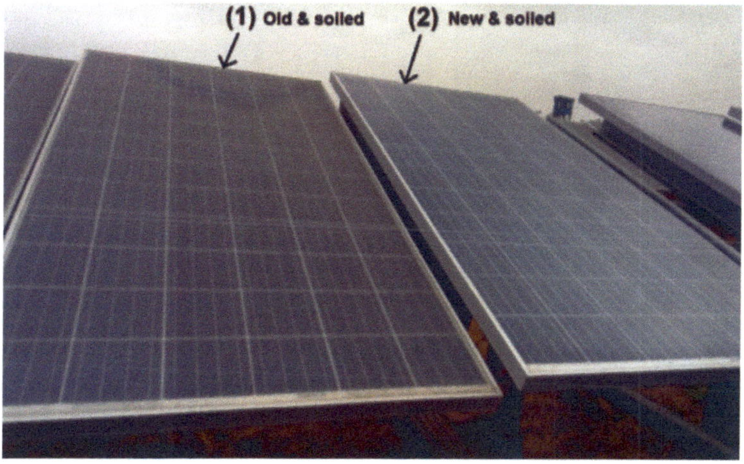

Figure 7.1 Solar panels under study.

Two panels are shown here: (i) The first one is old and soiled; and (ii) the other is new and therefore in a clean (without soiling) condition. During the study, the panels equipped with soiling measurement apparatus are allowed to be monitored continuously for 22 days, i.e., from 21 December 2018 to 11 January 2019. Soiling is allowed for both the panels under study. The new panel is cleaned every day with normal water during the test, whereas the old one can retain soiling. During the test, the soiling test kit is used to measure and store the data continuously. The kit consists of a clamp meter and a data logger. On the first day of the test, the output panel current's calibration is done using a clamp meter. Subsequently, the data logger is utilized to obtain continuous panel currents.

Soil particles also could be deposited on the uncleaned panel, which might extend surface conductivity. The decrease in surface resistivity due to soiling causes leakage currents on the front surface of the panel. The output currents and module temperatures are measured for each panel separately at the site. The humidity and ambient temperature data for the same time duration are obtained [27]. All the measurements are taken in a time interval of five minutes continuously for 22 days. After such an interval, measurements make it possible to investigate performance losses at an early stage of PID occurrence. Such data are collected for 287 data points per day, which leads to overall 6314 data points of panel currents. We neglect the effect of irradiation angle and soiling distribution. As described in Fig. 7.2, the suitable PV panels are selected first for the test after certain visual inspections.

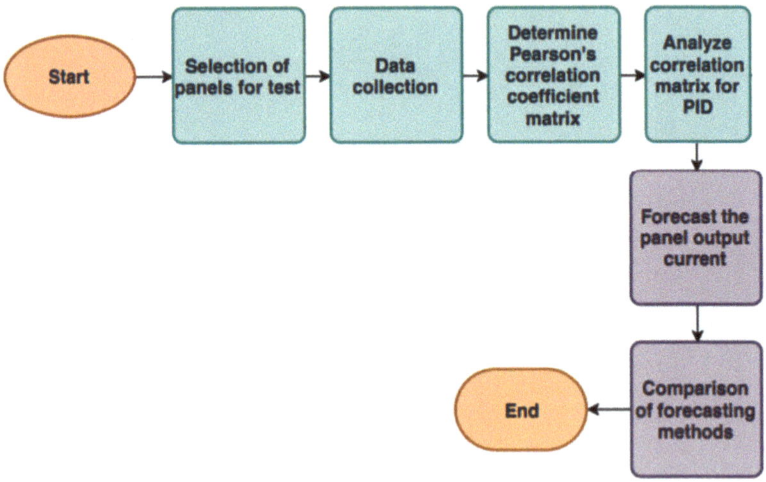

Figure 7.2 The process flow of PID data based Forecasting.

PV panels considered for the work are certain to be physically normal before the test. One new and one old solar panel are allowed to have soiling on their front surfaces. The new panel is cleaned every day with potable water, whereas the old panel has a soiled surface. Data such as the panel output current, module temperature, and relative humidity are collected for 22 days every 5 minutes of every day. During data acquisition, the impact of variations in irradiation angles and soiling distribution are considered to be unimportant. The collected data is then studied and analyzed to form the specific correlation matrix. Correlations are calculated among and between the module temperature, ambient temperature, relative humidity, soiled old panel current, and cleaned new panel current to form the matrix.

This matrix consists of Pearson's coefficients as the elements. Analysis of the Pearson's correlation-coefficient matrix is done taking the PID performance parameters into account. After an in-depth study of the correlation of other parameters with PID, the panel currents for the old and soiled solar panel are forecast. Various forecasting techniques in usage are PLA (Persistence Learning Algorithm), ARIMA (AutoRegressive Integrated Moving Average), SES (Single Exponential Smoothing), ANN (Artificial Neural network), SVR (Support Vector Regression), RF (Random Forests), and GBM (Gradient boosted machine). The ARIMA and SES methods are framed by using SPSS tool as in [28]. The accuracy and performance parameters of forecasting models are also studied. Lastly, a comparison between predicted outputs by all forecasting methods and actual data is studied to consider the accuracy and performance.

7.3 Forecasting methods

Basically, for any accurate forecasting method, the performance at different sites with varied atmospheric conditions is crucial. Electric power frameworks comprise numerous complicated factors with nonlear traits or attributes for which regression and time-series methods do not provide fitting outcomes. Algorithmic (machine learning-based) models can handle such complexities effectively. A parameter must be assigned that calculates how accurate the method is and determines fitness over certain conditions. Such optimization studies help choose precise and optimal forecasting methods such that the performance parameters evaluate and compare forecasting methods. The following subsections explain the schemes adopted depending on the prediction horizon. Also, we look at the various performance evaluation parameters.

7.3.1 Persistence learning algorithm (PLA)

This algorithm is the most common baseline method for supervised machine learning and also the most fundamental approach for forecasting the panel current. Previously, it has been used to predict several renewable energy attributes as in [29]. The prediction of a PID affected panel currently is made at $t + n$ by calculating the average of the actual soiled panel current for n previous forecast windows. The value of n is defined as per the requirement. This method is simple but not very precise because solar irradiance does change often quite drastically, and merely averaging past data is not very appropriate. However, it can serve as a good baseline method to compare other methods. One possible way of improving it is by detrending the data into a trend component and a random component.

7.3.2 Auto-regressive integrated moving average (ARIMA)

In 1970, Box and Jenkins formulated the ARIMA model, a straightforward statistical method utilized for time-series forecasting and portrayed by estimations of p, d and q, where p is the order of Auto-Regression (AR), d is the level of distinction to construct the model stationary, and q is the level of Moving-Regression (MA) [30]. For the values $p = d = q = 1$, the ARIMA method is applied for predicting the future values in this work. For the prediction intervals to be appropriate, the ARIMA models rely on uncorrelated residuals and normal distribution. For this reason, one ought to check the validity of the assumptions. Like most prediction interval calculations, ARIMA-based intervals tend to be narrow. It only accounts for the error variations, ignoring the parameter estimate variations and the model's order. The analysis assumes that the historical patterns in the model will continue during the forecast period.

7.3.3 Single exponential smoothing (SES)

This model comprises a constant smoothing value that places weight on the most recent value as in [31]. Additionally, the earlier period's most recent predicted value is likewise drawn to produce the new expected value in a traditional SES model. The application of every exponential smoothing method requires choosing the smoothing parameters and the initial values, and then the forecasts can be computed from the data. This process implements exponentially diminishing weightage to the previous values to predict the future values. For the overall model, the weightage can be selected utilizing a few strategies through surveying the correlation of the information and evaluating the weightage using the number of periods for this model built upon moving average (MA) idea, or through surveying the Residual Squared error (RSE) of a time series. The execution of the SES model for solar-panel current prediction in this paper is an updated one that only incorporates the past values without using the past prediction content to diminish the error caused by previously predicted readings.

7.3.4 Artificial neural networks (ANN)

In terms of forecasting, various machine learning algorithms like ANN have been developed to solve the problem of nonlinearity in prediction models [32]. Given any data set, ANN learns from the experience of the past like the human brain. The model is built based on past data, which is then utilized to predict future values. An artificial neural network's basic

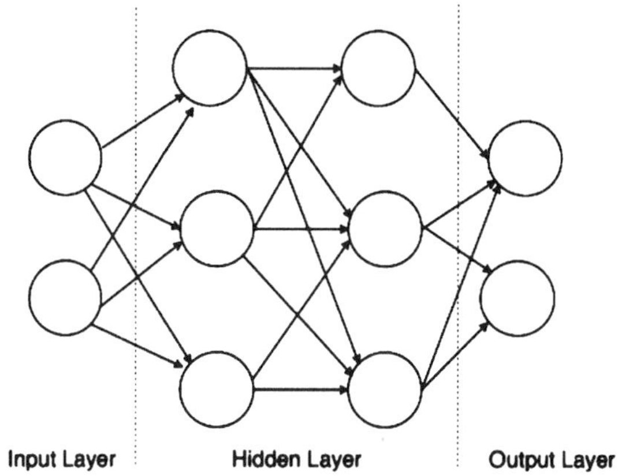

Figure 7.3 Multilayer ANN architecture.

operation is based on the brain's working, i.e., communication via neurons. Fig. 7.3 shows the multi-layer architecture for an ANN. Generally, it consists of an input layer, output layer, and hidden layer(s). The input layer of ANN has a weight associated with it. For desired output, the weights are updated in the model:

$$f(x) = \phi\left(\sum_{i=1}^{N} w_i x_i + b\right) = \phi\left(w^T x + b\right). \tag{7.1}$$

For (7.1), w_i are the weights associated with each neuron present in the input layer, x_i are the inputs, and b is the bias term. The function ϕ denotes the activation or transfer function (here sigmoid function). The neurons' weighted sum output gets sent to the activation function block, where a mathematical function scales the output to zero or one. Several combinations of ANN architectures are possible by keeping a different number of hidden layers and different number of neurons in each layer, along with the choice of activation functions. Next, we will try to answer a basic question, what is machine learning? A machine learning method is a computer algorithm that searches for patterns in data.

7.3.5 Support vector regression (SVR)

Normally, Support Vector Machines (SVM) utilize linear models to execute nonlinear class models through some nonlinear mapping of input vectors x into the high-dimensional element space. Formerly, it has been used for forecasting of solar output parameters [33]. It is basically used for tackling machine learning classification applications. The kernel trick is used for converting the classes into higher dimension to make the classes linearly separable after incorporating binary classes in 2-D space. The possible principal kernels are linear, polynomial, radial basis function. What is a hyperplane? How to classify high-dimensional data using an analytical expression with no visualization hints? How to assign classes to the objects of interest? The answer to these questions are critical to the application of SVRs.

The SVR model carries out the properties from SVM. There exists several kernels for yielding the inner products to build machines with various kinds of nonlinear decision surfaces in the given input space. Among kernels, the model that minimizes the estimate is the most suitable one. In SVR, the classification is carried out between the regression errors that are more or less than a certain threshold value. For SVR, the following objective function needs to be solved:

$$\min_{w,b,\xi} \frac{1}{2} W^T W + C \sum_{i=1}^{l} \xi_i, \tag{7.2}$$

$$\text{subject to } y_i \left(W^T \phi(x_i) + b \right) \geq 1 - \xi_i. \tag{7.3}$$

The $(x_i,\ y_i)$ training-data sets vary for the value of i=1 to l. W is a normal unit vector, which acts perpendicular to boundary margin, b is the value of slack variable, ϵ is the threshold parameter, $C > 0$ is the penalty parameter, and ξ is the deviation larger than ϵ.

Historical data in machine learning is a boon for data analysts around the world. In terms of regression models, benchmark models like persistence algorithm and ARIMA fail to express the nonlinear trend in wind speed time-series. Machine learning models with kernel mapping capability are widely applicable in such cases. In terms of accuracy, machine learning models based on supervised learning have good generalization performance, which is the ability of a machine learning model to adapt to an unforeseen data. The need for ensemble-based forecasts arises from the problem of over-fitting that occurs in models like support vector regression, multiple linear regression, and neural networks.

We discuss the over-fitting scenarios in the subsequent chapters considering real-time datasets. Decision trees aim to arrive at a decision which can be of two types: a continuous one which is called regression and a discrete one called classification. The fundamental problem is we have data that separates into two or more classes, and we're interested in finding a machine learning object or a mathematical function called a classifier that predicts which class a data point belongs to.

7.3.6 Random forests

Random Forests (RF) comprise a combinatorial classifier and can be utilized for regression as well. The fundamental purpose of RF is to formulate several decision trees during the training time and yield outcomes produced from individual trees [34]. An RF forest begins with splitting the input features into a group of subsets that essentially form a tree. A particular tree is characterized by a node that leads to number of branches. Similar to hyperparameters (σ and C) in SVR, in random forest regression, the number of trees and the number of random features in each tree decomposition are the parameters that decide the performance of regression, as depicted in Fig. 7.4. At each decision tree, a fitting function is created that acts on the random features selected. Finally, at the end of the training process, a random forest model is created. During the training process, each tree is created from randomly selected input vectors and so is called a 'random' forest.

RF depends on a choice tree utilizing an irregular choice of qualities for every choice-tree node. Each decision tree of the RF is prepared to use random subsets of the dataset. All subsets are constructed by selecting data arbitrarily from the originally fed dataset. The number of decision trees is determined ahead of time, and the information regarding the training dataset is collected. Mathematically, the estimated output of a random-forest regression is given as

$$\hat{A} = \frac{1}{k}\sum_{i=1}^{k}\hat{r}\left(X, V_i\right), \tag{7.4}$$

where $\hat{r}\left(X, V\right)$ is the representative tree at the end of training process, X is the set of input feature vectors, and T is the collective set representing the input–output pair $V_i = (x_1, y_1), (x_2, y_2), \ldots (x_n, y_n)$. The predicted output is averaged over k decision trees. An added advantage of random-forest regression is its insensitivity to noise due to uncorrelated trees via differential sampling of inputs. Fig. 7.4 illustrates the block diagram of a random-forest regression model. Decision trees are sensitive to the data through which they are trained. Changing the training data can change the predictions.

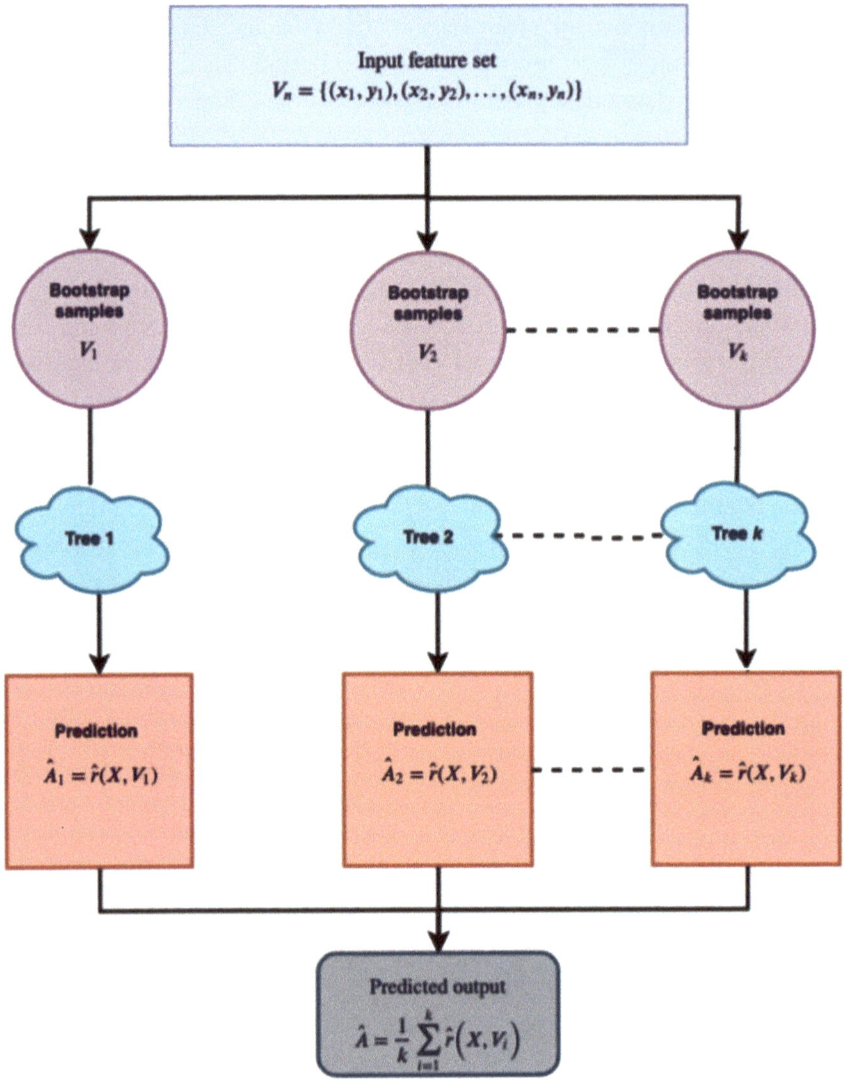

Figure 7.4 Random-forest algorithm.

Meanwhile, a common problem of over-fitting persists in machine-learning regression models when a well trained model captures the noise component as well. To reduce the complexities posed by over-fitting, a random forest compromises between a flexible and an inflexible model. In the training phase, each regression tree draws a sample from the feature set and is drawn repeatedly. This ensures that, even though the tree might possess a high variance, the overall variance of forest will be low. Random forests work on the principle of bagging, which combines the predictions from different tree models to give an overall insight about the data under training. This also helps to reduce the potential over-fitting caused by supervised machine-learning models. In case of bagging, the principle of negative correlation learning (NCL) is also applicable for regression-based problems [35]. In NCL, a penalty term is added while computing the error of an individual sample in an ensemble. The modified error function for an individual sample i is

$$E_i = \frac{1}{N}\sum_{i=1}^{N}\left(F_i(X) - y_i\right)^2 + \frac{1}{N}\sum_{i=1}^{N}\beta_p p_i, \tag{7.5}$$

where $F_i(X)$ is the predicted output for i^{th} sample, y_i is the actual value, $\beta_p \in [0,1]$ is a term used to control the strength of the penalty, and p_i is the correlation function.

The RF prediction process takes the mean of the predictions made by the individual regression trees. Then, numerous decision trees are developed furthermore. Ultimately, it gets the characterization results by a voting method that upgrades the classification implementation. Following the steps, RF regularly achieves the preferred exactness over every decision tree and can manage substantial datasets. The random-forest technique's significant parameters are tree depth and the number of less sensitive trees and can be effectively used for the current panel forecast. To get optimal results, the number of decision trees ranges from 100 to 1000, and this value is taken to be 500 in this case.

7.3.7 Gradient boosting method

The Gradient boosted machine (GBM) is a less often utilized machine learning approach for solar estimations [36]. It is, in some sense, an expanded version of the random-forest method. In RF, an arbitrarily assembled group of decision trees is utilized to make an increasingly exact estimation. This technique gradually adds trees to the group during the training stage. The GBM method is used for converting weak learners into strong learners by making certain additions of trees to the original data set. Every newly obtained regression tree is included in the ensemble portion one at a time to correct the residuals calculated earlier. The residuals are nothing but the negative gradients that measure the inconstancy level between an estimated and the normal result. Including trees in each one, in turn, enables each new tree to be explicitly prepared to improve the already ready outfit, instead of the RF procedure of arbitrarily adding nodes to new trees to accommodate a superior average. Therefore, by the inclusion of more regression trees, the accuracy is improved above a particular level.

Gradient boosted machines (GBM) is an ensemble-based regression method that takes into account the loss of the previously fit decision tree. In a GBM model, various weak learners combine with each other to arrive at an accurate model. Boosting-based regression trees are highly powerful for predicting accurate response values owing to the advantages posed by an appropriate boosting algorithm. A particular boosting algorithm strengthens the tree model by minimizing the inaccuracies of weak models in the form of a loss function [37]. Thus, for a boosting algorithm, the main task is to reduce the error posed by weak learners in each iteration. The final predicted response is the linear combination of fit trees. An advantage of GBM over conventional regression models is that it avoids over-fitting. The learning rate of each decision tree determines the predictive performance. A higher learning rate indicates that a lower number of trees are needed for prediction and vice-versa [38].

Based on a cross-validation experiment, the ideal number of trees required along with learning rate is obtained. Since GBM is a collective model that combines all the individual weak models, the aim of an individual model is to reduce the error generated by a loss function. Commonly, $\mathcal{L}_2$ loss function is used to minimizes the sum of squared errors between the predicted and actual values. Consider k such weak models that account for prediction of a variable y given feature set $x = [x_1, x_2, \ldots, x_n]$, which can be given as

$$\hat{y} = \sum_{i=1}^{k} f_i(x), \tag{7.6}$$

where $f_i(x)$ represents each weak learner that is collectively trained to improve the prediction results. The loss function is given by

$$\mathcal{L}_2 = \frac{1}{N} L(y_i, \hat{y}_i), \quad L(y_i, \hat{y}_i) = \sum_{i=1}^{N} (y_i - \hat{y}_i)^2, \tag{7.7}$$

where $L(y_i, \hat{y}_i)$ is a loss function based on squared errors for N observations. The aim of a GBM regression technique is to minimize the $\mathcal{L}_2$ loss function. However, the $\mathcal{L}_2$ loss function is more sensitive to outliers and can reduce the robustness of the model. To ensure high robustness, the Huber loss function is used, which can be expressed as

$$L(y, \hat{y}) = \begin{cases} \frac{1}{2}(y - \hat{y})^2, & |y - \hat{y}| \leq \delta \\ \delta(|y - \hat{y}| - \delta/2), & |y - \hat{y}| > \delta, \end{cases} \tag{7.8}$$

where δ represents the slope. The concept of GBM revolves around gradient descent and the boosting algorithm.

For optimizing the hyper-parameters in GBM, a gradient descent algorithm is used to minimize the cost function, taking into account the negative of the gradient. For a $\mathcal{L}_2$ loss function, the gradient with respect to predicted sample $\hat{y}_i$ is

$$\frac{\partial L(y, \hat{y})}{\partial \hat{y}_i} = \frac{\partial}{\partial \hat{y}_i} \sum_{i=1}^{N} (y_i - \hat{y}_i)^2 = 2(y_i - \hat{y}_i) \frac{\partial}{\partial \hat{y}_i} (y_i - \hat{y}_i) = -2(y_i - \hat{y}_i). \tag{7.9}$$

The gradient here reflects that, while tracking the minimum point of the loss function, the GBM actually tracks the residual vector $(y - \hat{y})$. Similarly, for $\mathcal{L}_1$, the MAE can be tracked by finding the gradient of loss function given as

$$\frac{\partial L(y, \hat{y})}{\partial \hat{y}_i} = \frac{\partial}{\partial \hat{y}_i} \sum_{i=1}^{N} |y_i - \hat{y}_i| = -sign(y_i - \hat{y}_i). \tag{7.10}$$

The accuracy of a GBM model is dependent on the number of weak learners and the learning rate. Further, the essence of a GBM model lies in the boosting technique that combines all the weak models in a stage-wise manner rather than in a parallel manner, as seen in random forest model. Mathematically, the GBM model in recursive form is given as

$$F_m(x) = F_{m-1}(x) + \eta \Delta_m(x), \tag{7.11}$$

where η is the learning rate and Δ_m refers to a regression model fitted to the residuals. Fig. 7.5 illustrates the *XBoost* algorithm for obtaining the best regressor.

Song et al. presented GBM model for predicting clathrate hydrate phase equilibria in salts [39]. The inputs to the model include 11 variables in terms of gases like CH_4, CO_2 and N_2, salts KCl and NaCl. Two models for deep neural network (DNN) and gradient-boosted regression trees with pressure and temperature as outputs are used. Results reveal that the GBM

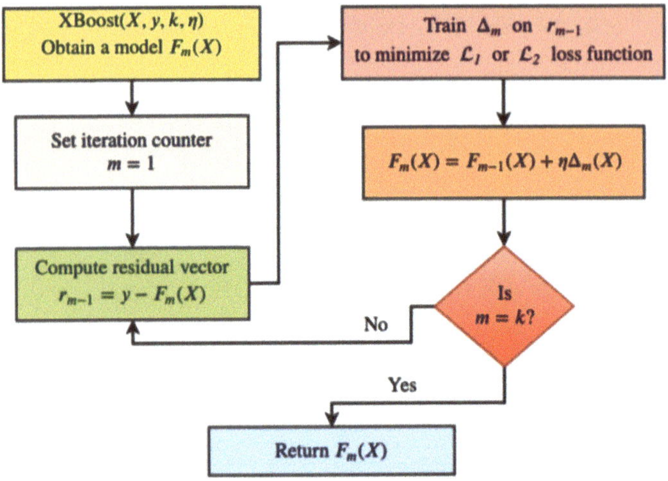

Figure 7.5 XBoost algorithm for gradient boosted machines.

model outperforms DNN for both temperature and pressure as inputs to the model. Further, the GBM model based on temperature as output gave better performance than the GBM model based on pressure. Miskell et al. discussed a GBM model for predicting the short-term peak concentration of $PM_{2.5}$ particles for Christchurch, New Zealand [40]. Variables like temperature, relative humidity, wind speed, wind gust, and air pressure are used as predictors in the forecast model.

7.4 Results and discussions

Daily variations in the panel currents indicate the correlation between soiling and PID. The effect of the irradiation angle is considered negligible for both panels. Fig. 7.1 shows that irradiation from the sun is consistent throughout the measurements. We observe the variations in the current yielded by the old, soiled (and new cleaned) panels alongside changes in atmospheric conditions. After the correlation study of the respective parameters, we forecast the panel currents using various forecasting techniques. A practical comparison of forecasting methods is made, based on the criteria of accuracy, to determine the optimum method for predicting the panel current and thereby the harmful effects of PID.

7.4.1 Correlation study of parameters

The correlation between the relevant parameters indicates a reduction in the panel current due to the PID phenomenon. For obtaining the correlation, in 22 days, 6314 data points for each parameter is collected. The parameters undergoing the correlation analysis are;

1. I_{soiled}: the magnitude of the soiled PV panel current in Amperes (A). The predicted dependent parameter uses various forecasting schemes.
2. $I_{cleaned}$: The magnitude of the cleaned PV panel current in A. It is considered an independent parameter during forecasting.
3. T_{mod}: the value of the temperature (in Kelvin (K)) of the PV modules inside a panel. The PV module temperature is considered to be approximately equal to the PV panel temperature.
4. *Rel.Humidity*: The value of relative humidity (in percentage (%)) for the test setup utilized in this work.

Fig. 7.6 describes the correlation between the considered parameters. The values of I_{soiled} and $I_{cleaned}$ are comparatively less. Module temperature and relative humidity variations can be significant.

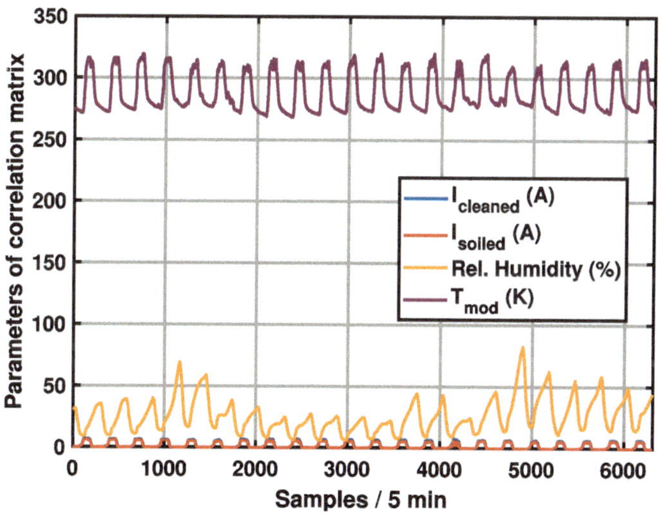

Figure 7.6 Correlation graph.

The parameters of correlation matrix, i.e., $I_{cleaned}$, I_{soiled}, Rel. Humidity, and T_{mod} are shown along with units, respectively. At the start and end of the day, we observe the value of soiled PV panel current to be zero due to no irradiance. Fig. 7.6 shows a direct positive correlation between the soiled and cleaned panel current and an inverse correlation between relative humidity and module temperature. The value of the soiled panel current is inversely proportional to the value of relative humidity. After analyzing 6314 data points of every parameter considered, the correlation between all the parameters forms a Pearson's correlation coefficient matrix.

The correlation matrix consists of Pearson's correlation coefficients as its respective elements. Pearson's correlation coefficient is a quantitative measure of the strength level of the correlation of two variables. It can vary from +1 to −1, wherein 0 indicates no direct correlation between the two parameters. A value more than zero (0) shows a direct positive relation, i.e., as one parameter increases, so do the other parameter. A value less than zero (0) shows a negative relation, i.e., as one parameter rises, another parameter falls. The Pearson correlation coefficient matrix is formed using the soiled old panel current (dependent parameter) and the independent parameters like relative humidity, module temperature, and the cleaned panel current. This matrix of order 4 × 4 is expressed as

$$\begin{array}{cccc} I_{soiled} & I_{cleaned} & T_{mod} & Rel.Humidity \end{array}$$
$$\begin{bmatrix} 1 & 0.998 & 0.934 & -0.419 \\ 0.998 & 1 & 0.938 & -0.417 \\ 0.934 & 0.938 & 1 & -0.287 \\ -0.419 & -0.417 & -0.287 & 1 \end{bmatrix} \begin{array}{l} I_{soiled} \\ I_{cleaned} \\ T_{mod} \\ Rel.Humidity \end{array}$$

As per the study done, a direct positive correlation between the soiled panel current and module temperature exists with the Pearson's coefficient value of 0.934, thereby confirming the proportionality of module temperature with the PID caused by leakage current. There is a direct inverse correlation between relative humidity and soiled panel current with the Pearson's coefficient value of 0.419, which substantially confirms the inverse relation of PID with relative humidity levels. There is a direct positive correlation between a soiled panel and a cleaned panel current with the Pearson's value of 0.998. Such correlation enables us to relate the effect of soiling to the occurrence of PID. Due to such leakage currents, overall module and panel currents are reduced by a significant amount. The leakage currents, which may be attributed to the effect of PID, act contradictorily to the panel currents in this case.

7.4.2 Forecasting results

To limit the occurrence of PID by the introduction of leakage currents, one must understand the overall behavior of panel currents with time. Forecasting methods can provide an overall reduction in the panel currents due to leakage. Accurate estimation yields the knowledge of the introduction and selection of suitable PID-recovery methods in advance. Tackling the PID by an appropriate recovery method empowers the users to extend a solar panel's life. This approach can conserve many resources and efforts to reduce the overall losses for a solar farm.

For forecasting and selecting the optimum technique, certain parameters are indispensable. Given the prediction horizon, the error between the forecasted value and the actual value of the soiled panel current can be quantified and compared using standard error definitions such as Mean Squared Error (MSE) and Root Mean Square Error (RMSE). These errors are calculated to find the accuracy levels of forecasting methods. Calculation and comparison of such parameters provide an optimum forecasting approach. The expressions for the MSE and RMSE are

$$MSE = \frac{1}{N}\sum_{t=1}^{N}(i_f - i)^2, \tag{7.12}$$

$$RMSE = \sqrt{\frac{1}{N}\sum_{t=1}^{N}(i_f - i)^2}, \tag{7.13}$$

where N is the prediction horizon, i_f is the forecasted current, and i is the actual panel current. The RMSE criterion for statistically analyzing the forecast accuracy has proven to be a better parameter. RMSE is crucial for the analysis and comparison of the forecasting methods. The least value of RMSE is the most desired one for determining the optimal forecasting scheme.

The soiled panel current is anticipated by incorporating several forecasting schemes and is compared with the actual one shown in Fig. 7.7. Seven different methods of forecasting the soiled PV panel current are implemented for a detailed study. The data is taken for a time interval of every five 5 minutes over 22 days. Therefore, 287 data points were collected for each parameter per day. Considering the 20% as the forecasting window, i.e., 1263 data points from the total 6314 data points, the validation is done for several forecasting techniques.

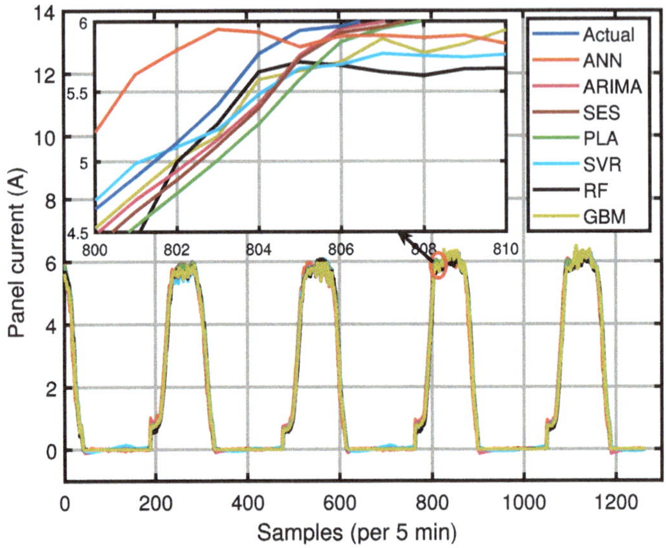

Figure 7.7 Forecasted data.

The variations of the predicted values of panel currents by different forecasting techniques are represented. To make it more appropriately visible, this is also displayed in the same figure expanded view. The expanded view consists of the samples from 800 to 810 data points. Furthermore, in the expanded view, the key differences in the predicted outputs can be observed. The distinctness and variations for predicted panel currents are displayed well in the expanded view. The predicted outputs are observed to calculate the preciseness of the methods adopted. It can be noticed in Fig. 7.7 that, at the start of the day, the uncertainties were observed in the predicted values of soiled PV panel current. Such uncertainties indicate that the starting irradiance level and time vary for different days. With climate changes and the corresponding effect on the ambient parameters, such variations are observed. It can be found that the predicted outputs obtained by using Persistence learning algorithm, are distinct from the actual values. Certain irregularities are also observed during the forecasting operations.

The maximum value of panel current is obtained as 6.3A predicted by the GBM algorithm. It can be clearly noticed that among all the forecasting schemes employed, the Gradient boosted machine (GBM) technique is found to be the most optimal. The output PV panel current predicted by using the GBM method is nearest to the actual PV panel current. The forecasted panel current by GBM method is within the reach of the actual panel current. As per the suitability, forecasting methods are framed either by utilizing SPSS or MATLAB software. The ARIMA and SES forecasting methods are applied using the SPSS tool, whereas the ANN, RF, PLA, SVR, and GBM methods are employed using MATLAB software. The relevant calculations and plotting of curves were performed using MATLAB software.

Table 7.1 Parameters for various time-forecasting techniques.

	ARIMA	ANN	PLA	
MSE	0.0106	0.0447	0.0463	
RMSE	0.1030	0.2114	0.2152	
	SVR	**RF**	**SES**	**GBM**
MSE	0.0095	0.0169	0.0150	0.0073
RMSE	0.0974	0.1299	0.1223	0.0855

Table 7.1 shows that the performance parameters for the various forecasting methods differ by significant magnitudes. Machine learning-based state-of-the-art forecasting techniques such as SVR, RF, and GBM are applied in this work. The major standard error definitions such as MSE and RMSE are calculated for every forecasting method. The least value of RMSE of 0.0855(8.55%) is found for the GBM technique. The highest value of RMSE is found for the PLA method, with an enormous value of 0.2152(21.52%). Other forecasting approaches, such as SVR, ARIMA, ANN, RF and SES, result in the RMSE values of 0.0974(9.74%), 0.1030(10.30%), 0.2114(21.14%), 0.1299(12.99%) and 0.1223(12.23%). The GBM technique is the optimal forecasting approach for panel current with the least value of RMSE of 0.0855(8.55%). If there are lower chances to incorporate GBM, then the SVR method has the least RMSE value of 0.0974 (9.74%) and can be deployed.

7.5 Conclusions

Numerous techniques and methods of forecasting are accessible to evaluate the panel currents. Some are regularly utilized (ANN, ARIMA, PLA), and others are rarely used (SVR, SES, Random Forest, and Gradient boosted machine (GBM)). At certain times, one is the best option, but it might be the case that a different method is best other times. By and large, the precision of these techniques relies upon the nature and size of the training data sets. Foreseeing solar panel currents ahead of time is a specific and efficient method for overseeing power demand and supply, especially where sustainable power-source innovations are a part of the portfolio. Although the adaptability of ANN to include nonlinear estimation makes it a better one than traditional ARIMA, the SVR method yields the lesser RMSE value of 0.0974 (9.74%). The Gradient boosted machine (GBM) leads to the least value of RMSE of 0.0855 (8.55%), making it the optimal forecasting method among all the implemented techniques. Adopting other state-of-the-art forecasting methods for PV panel currents can be considered under this study's future scope. Accurate early estimation of panel currents can assist in representing the reduction in panel currents due to the effect of PID comprehensively. Forecasting the panel attributes precisely helps in adopting optimal PID recovery and correction measures for limiting further contamination of panels due to the impact of PID. It may also assist in scheduling the PID detection tests. Furthermore, we can extend the study after implementing more advanced forecasting algorithms to protect the panels from the detrimental effects of PID.

Bibliography

[1] H.S. Dhiman, D. Deb, V. Muresan, V.E. Balas, Wake management in wind farms: an adaptive control approach, Energies 12 (7) (2019).

[2] J. Hattendorf, R. Low, W.-M. Gnehr, L. Wulff, M. Koekten, D. Koshnicharov, A. Blauaermel, J. Esquivel, Potential induced degradation in mono-crystalline silicon based modules: an acceleration model.

[3] P. Hacke, S. Spataru, K. Terwilliger, G. Perrin, S. Glick, S. Kurtz, J. Wohlgemuth, Accelerated testing and modeling of potential-induced degradation as a function of temperature and relative humidity, IEEE Journal of Photovoltaics 5 (6) (2015) 1549–1553.

[4] P. Hacke, K. Terwilliger, S. Glick, G. Tamizhmani, S. Tatapudi, C. Stark, S. Koch, T. Weber, J. Berghold, S. Hoffmann, M. Koehl, S. Dietrich, M. Ebert, G. Mathiak, Interlaboratory study to determine repeatability of the damp-heat test method for potential-induced degradation and polarization in crystalline silicon photovoltaic modules, IEEE Journal of Photovoltaics 5 (1) (2015) 94–101.

[5] P. Hacke, K. Terwilliger, R. Smith, S. Glick, J. Pankow, M. Kempe, S.K.I. Bennett, M. Kloos, System voltage potential-induced degradation mechanisms in PV modules and methods for test, in: 2011 37th IEEE Photovoltaic Specialists Conference, 2011, pp. 000814–000820.

[6] M. Schutze, M. Junghanel, M.B. Koentopp, S. Cwikla, S. Friedrich, J.W. Muller, P. Wawer, Laboratory study of potential induced degradation of silicon photovoltaic modules, in: 2011 37th IEEE Photovoltaic Specialists Conference, 2011, pp. 000821–000826.

[7] S. Pingel, O. Frank, M. Winkler, S. Daryan, T. Geipel, H. Hoehne, J. Berghold, Potential induced degradation of solar cells and panels, in: 2010 35th IEEE Photovoltaic Specialists Conference, 2010, pp. 002817–002822.

[8] P. Hacke, Characterization of multicrystalline silicon modules with system bias voltage applied in damp heat, in: 25th EUPVSEC, 2010, pp. 3760–3765.

[9] M. Schutze, M. Junghanel, O. Friedrichs, R. Wichtendahl, M. Scherff, J. Mueller, P. Wawer, Investigations of potential induced degradation of silicon photovoltaic modules, in: 26th EU PVSEC, 2011, pp. 3097–3102.

[10] D. Deb, N.L. Brahmbhatt, Review of yield increase of solar panels through soiling prevention, and a proposed water-free automated cleaning solution, Renewable and Sustainable Energy Reviews 82 (2018) 3306–3313.

[11] S. Mishra, K. Bhargava, D. Deb, Numerical simulation of potential induced degradation (pid) in different thin-film solar cells using scaps-1d, Solar Energy 188 (2019) 353–360.

[12] B. Espinar, J.-L. Aznarte, R. Girard, A.M. Moussa, G. Kariniotakis, Photovoltaic forecasting: a state of the art, in: 5th European PV-Hybrid and Mini-Grid Conference, OTTI - Ostbayerisches Technologie-Transfer-Institut, Tarragona, Spain, ISBN 978-3-941785-15-1, 2010, pp. 250–255.

[13] V. Lara-Fanego, J. Ruiz-Arias, D. Pozo-Vázquez, F. Santos-Alamillos, J. Tovar-Pescador, Evaluation of the WRF model solar irradiance forecasts in Andalusia (southern Spain), Solar Energy 86 (8) (2012) 2200–2217, progress in Solar Energy 3.

[14] A. Moreno-Munoz, J.J. de la Rosa, R. Posadillo, F. Bellido, Very short term forecasting of solar radiation, 2008, pp. 1–5.

[15] D. Anderson, M. Leach, Harvesting and redistributing renewable energy: on the role of gas and electricity grids to overcome intermittency through the generation and storage of hydrogen, Energy Policy 32 (14) (2004) 1603–1614.

[16] T.R. Ayodele, A.S.O. Ogunjuyigbe, C.G. Monyei, On the global solar radiation prediction methods, Journal of Renewable and Sustainable Energy 8 (2) (2016) 023702.

[17] C. Osterwald, T. McMahon, J. del Cueto, Electrochemical corrosion of sno2:f transparent conducting layers in thin-film photovoltaic modules, Solar Energy Materials and Solar Cells 79 (2003) 21–33.

[18] P. Hacke, P. Burton, A. Hendrickson, S. Spataru, S. Glick, K. Terwilliger, Effects of photovoltaic module soiling on glass surface resistance and potential-induced degradation, 2015, pp. 1–4.

[19] S. Pingel, S. Janke, J. Seydewitz, R. Alam, S. Koch, J. Kupke, J. Berghold, The local potential distribution as driver of pid & "live pid monitoring" method, 2014.

[20] N. Shiradkar, E. Schneller, N.G. Dhere, Finite element analysis based model to study the electric field distribution and leakage current in PV modules under high voltage bias 8825 (2013) 88250G.

[21] S. Hoffmann, M. Koehl, Effect of humidity and temperature on the potential-induced degradation, Progress in Photovoltaics: Research and Applications 22 (2) (2014) 173–179.

[22] R.H. Inman, H.T. Pedro, C.F. Coimbra, Solar forecasting methods for renewable energy integration, Progress in Energy and Combustion Science 39 (6) (2013) 535–576.

[23] J.R. Andrade, R.J. Bessa, Improving renewable energy forecasting with a grid of numerical weather predictions, IEEE Transactions on Sustainable Energy 8 (4) (2017) 1571–1580.

[24] Rec peak energy 72 – www.recgroup.com, https://www.recgroup.com/en/products/rec-peak-energy-72-family-powerplant, 2019. (Accessed 2 April 2019).

[25] P. Hacke, S. Johnston, All about pid-testing and avoidance in the field, Photovoltaics International 33, NREL/JA–5J00–NREL/JA-5J66799.

[26] J. Huang, H. Li, Y. Sun, H. Wang, H. Yang, Investigation on potential-induced degradation in a 50 MWp crystalline silicon photovoltaic power plant, International Journal of Photoenergy 2018 (2018) 1–7.

[27] Merra – www.soda-pro.com, http://www.soda-pro.com/web-services/meteo-data/merra, 2019. (Accessed 1 April 2019).

[28] R. Perdomo, E. Banguero, G. Gordillo, Statistical modeling for global solar radiation forecasting in Bogotá, in: 2010 35th IEEE Photovoltaic Specialists Conference, 2010, pp. 002374–002379.
[29] P. Patel, A. Shandilya, D. Deb, Optimized hybrid wind power generation with forecasting algorithms and battery life considerations, in: 2017 IEEE Power and Energy Conference at Illinois (PECI), 2017, pp. 1–6.
[30] I. Colak, M. Yesilbudak, N. Genc, R. Bayindir, Multi-period prediction of solar radiation using arma and arima models, in: 2015 IEEE 14th International Conference on Machine Learning and Applications (ICMLA), 2015, pp. 1045–1049.
[31] P.Y. Lim, F. Wong, Photovoltaic power predictions using modified adaptive response rate exponential smoothing method, in: 2016 IEEE International Conference on Sustainable Energy Technologies (ICSET), 2016, pp. 369–373.
[32] M. Narvekar, P. Fargose, Daily weather forecasting using artificial neural network, International Journal of Computer Applications 121 (2015) 9–13.
[33] A. Fentis, L. Bahatti, M. Mestari, B. Chouri, Short-term solar power forecasting using support vector regression and feed-forward nn, in: 2017 15th IEEE International New Circuits and Systems Conference (NEWCAS), 2017, pp. 405–408.
[34] C.-F. Yen, H.-Y. Hsieh, K.-W. Su, M.-C. Yu, J.-S. Leu, Solar power prediction via support vector machine and random forest, E3S Web of Conferences 69 (2018) 01004.
[35] Y. Liu, X. Yao, Ensemble learning via negative correlation, Neural Networks 12 (10) (1999) 1399–1404.
[36] J. Friedman, Greedy function approximation: a gradient boosting machine, The Annals of Statistics 29 (2001) 1189–1232.
[37] D. Solomatine, D. Shrestha, AdaBoost.RT: a boosting algorithm for regression problems, in: 2004 IEEE International Joint Conference on Neural Networks (IEEE Cat. No. 04CH37541), IEEE, 2004.
[38] J. Elith, J.R. Leathwick, T. Hastie, A working guide to boosted regression trees, Journal of Animal Ecology 77 (4) (2008) 802–813, https://doi.org/10.1111/j.1365-2656.2008.01390.x.
[39] Y. Song, H. Zhou, P. Wang, M. Yang, Prediction of clathrate hydrate phase equilibria using gradient boosted regression trees and deep neural networks, Journal of Chemical Thermodynamics 135 (2019) 86–96.
[40] G. Miskell, W. Pattinson, L. Weissert, D. Williams, Forecasting short-term peak concentrations from a network of air quality instruments measuring PM2.5 using boosted gradient machine models, Journal of Environmental Management 242 (2019) 56–64.

CHAPTER 8

Optimization of on-site PID detection methods

Solar cells comprise the most promising sustainable power source devices, causing tremendous interest in solar installations globally. Despite a huge demand, there is a deterioration of electrical attributes after some time in solar cells because of various internal and external parameters that require specific adjustments. Potential-induced degradation (PID) is one such phenomenon that unfavorably influences the electrical output parameters and impacts solar cells' reliability. Mainly, the effect of PID happens due to the introduction of leakage currents as a result of a combination of atmospheric attributes and soiling levels. In this chapter, the optimization of PID detection methods is treated for the solar panels in solar farms. Various PID detection methods are studied appropriately to suggest the optimal one. In particular, several Multi-Criteria Decision Making (MCDM) techniques are employed to find the optimal PID detection method among the existing ones. The performance scores of various MCDM techniques are compared and studied to identify an optimal PID detection method.

8.1 Introduction

With a significant increase in the establishment of sustainable power sources, fossil fuel usage is decreasing. Globally, the large majority of planned power installations exploit sustainable-power sources like solar and wind due to their lower carbon footprint, and they need to be managed by using advanced control techniques [1]. With solar energy abundantly available, tapping power efficiently from solar panels is a critical task. Due to such high demand, the crystalline-type solar cells' adaptability is advancing day by day. While solar modules are considered a useful resource, they also have certain limitations. A well-known degradation phenomenon is Potential-Induced Degradation (PID), occurring in solar cells due to the migration of sodium ions through the potential gradient from the front SLG substrate inside the anti-reflective coating (ARC) of a PV module. Numerous factors affect the pace of the onset of PID in the crystalline-type solar cells, like humidity and temperature levels [2,3], illumination levels [4], recov-

Degradation, Mitigation, and Forecasting Approaches in Thin Film Photovoltaics
https://doi.org/10.1016/B978-0-12-823483-9.00019-X

ery history [5,6], the voltage potential of the active cell circuit, and system grounding mechanisms [7].

A significant amount of PID growth happens when the surface layer electrically conducts, leading to the formation of a ground electrode against the negative bias of the cell [8,9]. The surface layer soiling is also known to have the harmful effect of increasing the PID on solar panels. Soiling absorbs the humidity like a sponge, which leads to persistent surface groundings due to high leakage currents. It is foreseen that the presence of soiling on module glass leads to diminished module surface resistance, further extending the potential across the glass surface of a grounded PV-module frame. Various control strategies and cleaning methods have been proposed so far to provide electrical output reliably from the solar panels [10]. Also, at the cell level, the effect of PID is studied for various thin-film solar cells by introducing defect states at specific layers [11]. In this volume, this study extends to the PV-system level to estimate optimal PID detection and recovery techniques.

Electrical-power personnel should ensure a balance between the power generation and utilization at any one time, and this is a challenge to maintain within the traditional and controllable energy production framework. These days, several nations are considering utilizing sustainable power sources in their power systems, which creates challenging issues when the source (wind, solar, etc.) is variable. It is essential to anticipate the solar-output data adequately for a highly coordinated energy system. The biggest challenge for the new era is the successful integration of renewable energy-power sources. The power system's overall reliability depends on the system's ability to account for the unexpected variations and interruptions, while ensuring quality and congruity of service to the clients.

The combination of renewable energy sources and an existing power network escalates the grid system's complexity and threatens the production/consumption continuity due to their irregular nature [12,13]. The uncertainty in the production of power by solar devices entails other obstacles like voltage glitches, instability, power grades, and integration issues [14,15]. Due to the gradual shift to renewable energy sources, the precise prediction of solar outputs becomes a primary concern for highly coordinated power frameworks [16]. In this work, panel current is forecast, which helps to make early estimates of the negative effect of PID. As a consequence, suitable recovery and correction measures are employable in advance.

Less progress has taken place to determine the correlation between the influence of PID and soiling levels. The conductivity of the surface of the active cell circuit over the glass front affects the process of PID due to the grounded paths [17]. Due to the conductivity of soil deposited on the panel, the surface resistance of module glass is reduced [18]. Surface soiling, which depends on the ambient parameters, causes increasing leakage currents at lower levels of surface humidity. For PV modules packaged with a relatively low front-face bulk resistivity, the ionic leakage extending across the module is ended due to a variety of front-surface lateral resistances [19]. Using materials with higher resistivity causes a more uniform distribution of leakage currents for a PID affected module [20]. The leakage currents are generated on the glass surface with an increased contact with the ground potential. The PID rate may immediately have a direct correlation with leakage current for a given PV module.

The leakage currents induced may lead to PID losses. Earlier, we explored the effect of temperature, relative humidity, bias voltage, and contact to the ground on the leakage current's value [21]. There exists an Arrhenius relation between temperature and leakage current for an activation energy magnitude of 75 $kJ\,mol^{-1}$. Also, an ohmic correlation exists between leakage current and bias voltage. The leakage current is lower for low surface humidity and higher with a contacting metal foil at the surface. As discussed earlier, the leakage current between the cells and frame also gets subjected to ambient moisture. There is a development of a thin layer on the surface glazing front which conducts electricity beyond a particular percolation threshold magnitude.

In this research work, we forecast the panel currents after conducting the soiling tests. Various forecasting models are used to anticipate future values. Because the panel currents decrease due to leakage currents related to PID, such investigations help assess the future variations in the panel currents due to PID's impact. The panel currents get subjected to PID losses because of such induced leakage currents are due to soiling. This chapter rigorously presents the ongoing and advanced forecasting plans for the panel currents in a PID-affected solar panel that can prompt early estimation of PID effects. Proper PID-recuperation measures are secured in favor of healthy solar panels. The investigation is intended to improve the economies of the size of the extensive sustainable-energy frameworks. The expanded awareness of solar PV panels as a sustainable power source has expanded the quantity of PV panel installations recently. Meanwhile, more accessibility of information and computational power have empowered machine learning calculations for improved forecasts. Because the need to foresee PV performance parameters is fundamental for the whole energy sector, forecasting techniques are important to understand the effect of PID on output panel currents. Several machine learning and statistical methods can be incorporated for solar forecasting [22]. In this examination, a correlation between various PV performance parameters and time-series forecasting models is undertaken for a solar farm. Proper studies and monitoring of solar cells are needed before incorporating them into the solar farm. Several factors affect the depth of PID in solar cells, mainly the ambient parameters such as temperature and relative humidity. The effects of ambient parameters need to be studied to know about PID dependence.

Numerous machine learning techniques are explored and evaluated in combination with various features meant to ameliorate the overall performance [23]. The time-series forecasting methods are somewhat complicated due to the nonstationary energy time series. We study the PV panel current's general behavior during the variation in the ambient parameters and the soiling level changes. Apart from yielding the optimal forecasting method for panel current forecasting, the study will also help obtain the correlation between PID and soiling levels depending on the wind speed and relative humidity. Other performance parameters can also be studied for a better understanding of PID. The PID here results from soiling level, temperature, and relative humidity.

In this work, the correlation between the studied parameters is made to determine the actual effect of PID. After the correlation study, various forecasting methods are utilized for further analysis. The work attempts to find the optimal forecasting technique for estimating the critical nature of PID on the PV panel currents.

8.2 Adopted methodology

PID became a critical issue years ago because the solar industry became concerned over its impacts as it could cause up to 80% losses. PID was originally identified primarily with high framework voltages, typical of the frameworks introduced after 2005. Such high framework voltages, when united with ambient parameters, facilitate the PID effect, as discussed. The ambient parameters, such as module temperature and relative humidity, affect the rate of PID occurrence. The PID detection becomes crucial for early knowledge of PID without causing much harm to the solar panels. As there is no strictly defined standard test system to detect PID, various

PID detection methods are incorporated based on the application type. The absence of any specific accepted measure of PID is the motivation behind the recommendation of several types of tests by industrialists and scholars for the PID recognition in labs [24]. During the study here, certain steps are followed in order to get the desired results. The flowchart of the work is shown in Fig. 8.1.

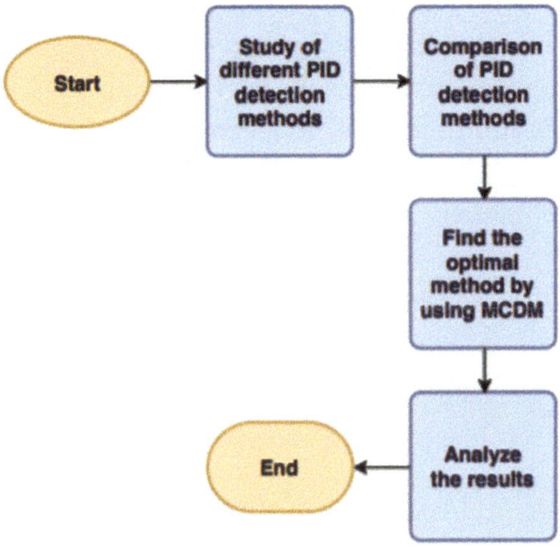

Figure 8.1 Flow chart for optimal method identification.

As observed in Fig. 8.1, initially, we study the existing methods for on-site detection of PID. This study is done as a part of the system-level study of PID. There are mainly six basic PID detection techniques that are considered. After the in-depth research, a comparison of such methods is made based on cost, reliability, and usability, which are the criteria on which the optimization is performed. MCDM techniques come into play after the detailed comparison of the PID detection techniques. Using the MCDM technique, the overall decision matrix is formed considering cost as a nonbeneficial criteria and others, i.e., usability and reliability, as beneficial criteria. By employing the MCDM technique, the optimal method for PID detection is selected. After that, the analysis of the obtained results leads a proposed optimal PID detection technique.

8.3 Existing PID detection techniques

With the increase in the number of solar panel frameworks day by day, there is a requirement for a better and in-depth understanding of the PV module's performance parameters. PID is a harmful phenomenon in which induction of leakage currents occurs due to the mobility of sodium ions from the SLG substrate to the frame. The effect of PID is found to be contagious and can increase after a specific time. There is a critical requirement of on-site detection of PID for a solar panel in a solar farm. Study and optimization of on-site PID detection techniques help conserve resources related to the PV framework. The proper selection of an on-site PID detection technique will also help perform the tests considering the criteria of cost, reliability, and usability. In this work, we first explain all the existing PID detection methods and their possible spectrum of applications. Apart from the operation's scope and the associated demerits, we also discuss several PID detection techniques.

8.3.1 Electroluminescence (EL) imaging

Electroluminescence imaging (EL) is a popular method that aims to ascertain electrical contacts' effectiveness in solar cells [25] by identifying cells' inherent defects like bad soldering, debris, and most importantly, micro cracks which can potentially and significantly affect performance. EL imaging takes advantage of the radiative interband recombination of excited charge carriers in solar cells. This method needs electrical contact and is only usable after completing the cell's metalization step when the cell is significantly complete. A moderate and definite check to find whether a PV module is PID affected or not is by getting an image of its EL test, accomplished by incorporating a charge-coupled device (CCD) camera which is an expensive device. It is performed only in the absence of daylight by biasing the PV module with a current source. This test can be done only on-site at night. A PID-resistant module has an EL picture with every cell possessing a nearly equivalent brightness level, but a module influenced by PID has somewhat dimmer cells due to the shunting of cells, as in Fig. 8.2.

As noticed in prior study [26], the PV modules nearer to the positive pole have a nearly similar brilliance. In contrast, the modules closer to the negative pole have dimmer cells that are PID influenced. This method doesn't explain how such degradation will affect the PV module output parameters. Despite higher reliability, it is not an economical method to detect PID due to to need to employ an expensive device. The conditions needed during the test make it less reliable. The EL test method needs at least two skilled people to perform it, which leads to its only moderate implementation.

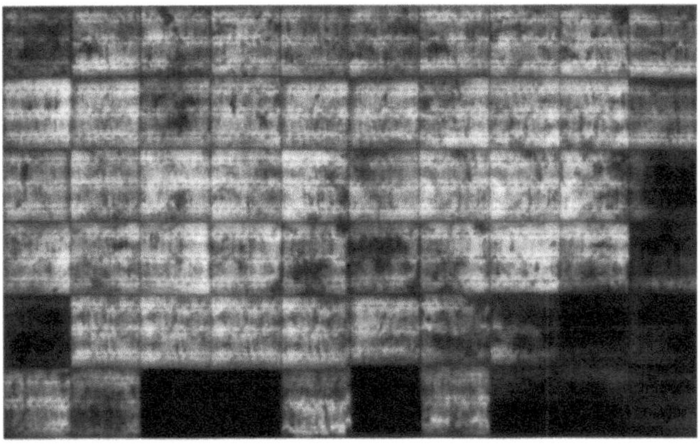

Figure 8.2 Electroluminescence (EL) imaging [26].

8.3.2 Thermal imaging

Thermal imaging refers to the usage of an Infrared (IR) camera to detect PID. Module review by using an IR camera can be a valuable technique to gauge whether a PV module is influenced by PID [27]. Uncooled, handheld, and lightweight thermal imaging cameras are currently in wide use; they result in a great deal of flexibility for solar-panel quality controls before installation and regular preemptive maintenance check-ups afterward. For adequate thermal contrast during the inspection of solar cells in the field, a minimum irradiance of 500 W/m^2 is necessary, and 700 W/m^2 is desirable, heavily depending on the location and local weather. Lower ambient temperatures may potentially raise the thermal contrast.

As seen in Fig. 8.3, the cells nearer to the positive or negative poles are at higher temperatures. A PID module can have cells at high temperatures, but a module free of PID has every cell approximately at the equivalent temperature. Also, these cells at higher temperature are at the positive (or negative) pole of the PV string, as can be seen in Fig. 8.3. This type of test can be effectively managed without meddling in the activity of the PV establishment. Not always does an IR investigation uncover the PID. Moreover, this IR review must be done on a bright day to identify the major differences between healthy and PID-affected modules appropriately. Considering the scope of the work, it can be implement on any solar farm during the daytime by a single skilled person. It may sometimes show inaccurate results due to particle deposition on the PV panel's surface. Repetitive measurements after cleaning the surface can be done to ensure the accuracy of the IR camera.

Figure 8.3 Thermal imaging [27].

8.3.3 Open-circuit voltage measurement

There are other options of quantifying PID rapidly by using a voltmeter to measure the open-circuit voltage (V_{oc}) of the PV module (see Fig. 8.4) [28]. It is measured by measuring the voltage across the open terminals of the PV-module. For a PV module influenced by PID, it is found that the value of V_{oc} can be lower than normal. It was learned studied earlier that the decrease of shunt resistance due to PID can lower the open-circuit voltage. Another option to utilize V_{OC} measurements is the comparison of V_{OC} of the modules on the '+' side of the string to the V_{OC} at the '−' side. The existence (onset) of PID is almost confirmed if a difference in these two voltages exists beyond a given threshold.

However, this decrease is possibly only seen when the impact of PID is high. As discussed earlier, the contrast between the V_{oc} of the normal and PID affected modules is small. Thus, the V_{oc} does not always uncover if PID influences a module because sometimes none is influenced. These measurements must also be done during the daytime and within in a brief time period to ensure that all the measurements are done at equivalent irradiance levels and ambient conditions with the least errors. The voltmeter is easier to operate, but the reliability for PID detection is quite low. Thus, the open-circuit voltage measurement for PID detection has less scope for application in a solar farm.

Figure 8.4 Open-circuit voltage measurement.

8.3.4 Operating voltage measurement

Every module's operating voltage can be immediately measured using a voltmeter [28]. This estimate gives considerably more data than the V_{oc} value used. PID increasingly influences the parameters of modules with a working voltage lower than the normal ones. Because the modules with PID have higher leakage currents, a lower voltage is needed to convey a similar current than the rest of the string's modules. Additionally, this reduction of working voltage is relative to the overall PID losses. A more straightforward technique to quantify the operating voltage is by opening the junction box to approach the panel's poles. The operational voltage estimation must also be done on a radiant day and at regular intervals to ensure that every one of the estimates is done at a similar irradiance level and cell temperature conditions to estimate the various modules.

8.3.5 Current voltage (IV) curve tracing

The standard test to know whether a module is PID resistant or not is to gauge its IV characteristics using an electronic tracer [29]. Just the shape of the IV curve can report the nearness of abnormalities in the behavior. It can be observed in Fig. 8.5 that, due to the detrimental effect of PID, the shape of the IV curve changes.

Be that as it may, the distinctions in the shape are not usually so apparent. At that point, we expect to analyze the intensity of various modules at

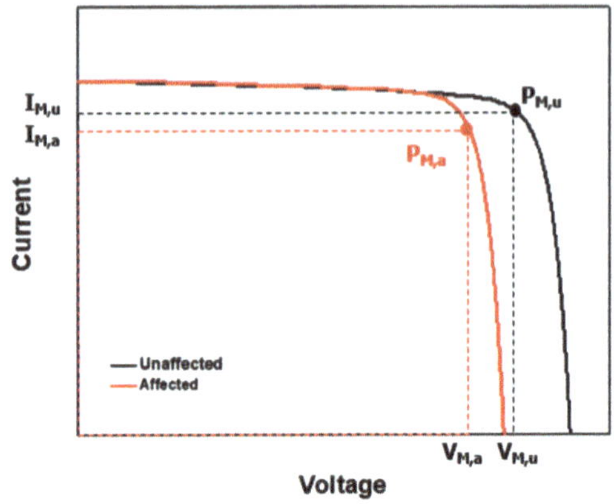

Figure 8.5 IV curve [29].

standard test conditions (STC). These outcomes are consistent with the estimations of the working voltage. Many specialists report that the output efficiency of PV modules at low irradiance is much affected by PID. In this way, another option is to quantify the IV curve of the PV module at low radiance levels. This should be possible by estimating the IV curve at the absolute top hours or bottom hours of the day. It must be done within a short time frame at the same irradiance levels to avoid any irregularities.

8.3.6 Dark current-voltage (IV) curve tracing

Finally, one more on-site test to identify PID is to measure the normal voltage of the PV module after biasing it with a particular current source during dark conditions [30]. This is often done with a cardboard or other opaque cover enclosing the entire PV module region, as shown in Fig. 8.6.

The estimations are clear with the trials of the working voltage and IV curve tracing. These centers address one of the motivations behind the dark IV method, typically for the modules. This method reports shunt resistances identified with the leakage currents due to the PID. In contrast to the past tests, this one has no time restrictions. The dark conditions that we need for this test are achievable with any covering, facilitating taking readings flexibly. As observed, the estimations are consistent with the previous operating voltage measurement tests and tracing of the IV curve. These focuses speak to one of the purposes of the dark IV regular for the modules. The dark IV curve investigates the introduction of leakage currents due to shunt resistances, leading to PID. This test has no time restrictions, so it's a much

Figure 8.6 Dark IV curve [30].

reliable and cost-effective method for detecting PID. It is flexible enough to be done at any point in time without affecting the modules.

8.4 Application of multiple-criteria decision-making (MCDM) techniques

Multiple-criteria decision-making (MCDM) or multiple-criteria decision analysis (MCDA) is a sub-discipline of operations research that explicitly simplifies multiple conflicting criteria in decision making for various fields [31]. Clashing criteria are ordinary in assessing alternatives, where reliability usually is one of the principal criteria. In this technologically advanced era, more often, people don't gauge numerous criteria and might be alright with the results of such choices that are merely dependent on intuition. Organizing complex issues well and considering various criteria unequivocally prompts progressively informed and other better choices. There have been significant advances in this field. There are several MCDM techniques as per the suitability and accuracy levels desired. Concerning this work, the MCDM techniques incorporated are Simple Additive Weighting (SAW), technique for the order of preference by similarity to the ideal solution (TOPSIS), and Elimination et Choix Traduisant la Realite (ELECTRE). The methods implemented here have their own operational principles and calculations. The methods implemented are explained in the following sections.

8.4.1 Simple additive weighting (SAW)

SAW is a weighted total technique. The SAW technique's fundamental idea is to scan for the weighted sums obtained from the scores of performance of every option concerning all considered criteria [32]. When implementing the SAW strategy, it is important to perform the final decision matrix's normalization on a scale equivalent to all choices. In the SAW technique, the three considered are cost, reliability, and usability. The basic steps involved in the process are

- **i)** Decide on the available choices.
- **ii)** Characterize the criteria that will be utilized as a source of perspective in basic decision making.
- **iii)** Make a performance rating table of every option on every criteria.
- **iv)** Determine the performance weights of options based on criteria.
- **v)** Create a decision matrix consisting all the performance weights.
- **vi)** Normalize the decision matrix after calculating the value of the normalized performance's performance.
- **vii)** Formulate the final normalized matrix.
- **viii)** The final step involves normalized matrix elements with the corresponding matrix weights across the decision matrix column.

8.4.2 Technique for order of preference by similarity to ideal solution (TOPSIS)

TOPSIS is a multi-criteria decision-making method that uses the basic mathematical approach to find the optimal alternative [32]. The guideline utilized in TOPSIS is that the chosen elective must have the nearest separate from the positive ideal solution and farthest from the negative ideal solution by incorporating the Euclidean distance to determine the relative proximity geometrically. The positive ideal solution ($X+$) is represented as the total of all best-obtained values for each alternative. In contrast, the negative ideal solution ($X-$) includes all the worst attainable values for each alternative considered. Both the solutions are hypothetical and are obtained within the process. This method does so by calculating the proximity related to the positive ideal solution. As per the relative comparison and calculations, we choose an alternative priority (Y). This MCDM method is widely used for solving real-life problems. With efficient computation and the caliber to calculate the relative performance of decision options, it is a straightforward technique.

8.4.3 Elimination Et Choix Traduisant la Realité (ELECTRE)

The Elimination Et Choix Traduisant la Realité (ELECTRE) strategy was considered by Bernard Roy [33]. ELECTRE isn't only a strategy but also an alternate choice that helps rationality. Not long ago, it was effectively connected to numerous fields. A solitary alternative or gathering of choices is chosen through a choice method and doled out to a kernel of viable options. The positioning of all alternatives considered in sets is accomplished by utilizing ELECTRE arrangement systems, relegating all choices to predefined classes. Edges and the outranking idea present the strategy. ELECTRE presents the edge thought so that the inclinations are characterized again. The critical decisions are the ones to illustrate the lack of changes in the threshold. Hypothetically there is a valid justification to embed a center zone in the middle of the lack of differences and the exacting inclination. ELECTRE produces an entire arrangement of twofold outranking relations among the choices.

8.5 Results and discussions

In this work, the overall decision matrix is calculated by considering the cost as a nonbeneficial criterion along with the reliability and usability as the beneficial ones. The PID detection methods are studied and compared using the defined criteria to calculate the decision matrix. The following table as a matrix is evaluated for finalizing the decision of the optimum PID-detection technique:

Cost	Reliability	Usability	
9	7	4	EL Imaging
8	5	5	Thermal Imaging
4	4	8	Open circuit voltage
3	3	9	Operating Voltage
5	8	7	IV curve
7	9	6	Dark IV curve

As observed in this matrix, all the PID detection methods are intensively studied for comparison based on cost, reliability, and usability. The weights are being calculated using the SAW method. The weights obtained by the SAW method are 0.3107, 0.3107, and 0.3786 for cost, reliability, and usability, respectively. As per the calculations performed, the various techniques for MCDM evaluation were incorporated, such as SAW, TOPSIS, and ELECTRE. Techniques such as SAW, TOPSIS, and ELECTRE are implemented using MATLAB®. After examining all the alternatives' performance scores concerning the considered criteria, the decision matrix is formed. The work is further extended to find the most optimal on-site PID-detection method after performing the MCDM techniques as explained.

Table 8.1 Performance parameters of MCDM techniques.

Alternatives	ELECTRE	SAW		TOPSIS	
	Rank	*Rank*	*Score*	*Rank*	*Score*
EL imaging	5	5	0.5134	5	0.3182
Thermal imaging	6	6	0.4994	6	0.2416
Open circuit voltage	3	3	0.7076	3	0.5683
Operating voltage	4	1	0.7928	2	0.5813
IV curve	1	2	0.7570	1	0.6936
Dark IV	2	4	0.6962	4	0.5598

As found in Table 8.1, it is clear that, by using the SAW method, the measured operating voltage was found optimal with the performance score of 0.7928 with the thermal imaging as the least optimal with the performance score of 0.4994. Similarly, by implementing the TOPSIS method, the IV curve technique was optimal for PID detection with a performance score of 0.6936. In contrast, the thermal-imaging approach was found least optimal, with a performance score of 0.2416.

The ELECTRE method gives the comparative analysis of the various techniques, finding that the IV curve technique is optimum for PID detection. The least effective approach was Electroluminescence (EL) imaging. The results reveal that the IV-curve tracing is the best on-site PID detection method in a solar farm based on the cost, reliability, and usability criteria by considering the options available and the data calculations. The proper and optimal PID detection technique's application depends upon the conditions and the criteria considered during usage. All the PID detection techniques studied so far have inevitable trade-offs for versatile utilization. Using the IV-tracing method, the PID detection can be done for a typical solar rooftop set-up. The IV tracing proves to be the most versatile PID detection technique for a quick, accurate, and cost-effective implementation.

Finally, the study explores the possibility of selecting an optimal method for detecting PID at the system level. After vigorous comparison of the existing techniques, the Multiple Criteria Decision Making (MCDM) method is implemented to select the optimal one. The results reveal that the IV-curve tracing is the best on-site PID detection method in solar farms based on the cost, reliability, and usability criteria by considering the options available and the data calculations. The study will help researchers and solar-cell manufacturers develop less PID-vulnerable solar cells, which would benefit humanity. At the same time, the appropriate selection of an optimal PID-detection method can save a great deal of resources and work hours.

Bibliography

[1] H.S. Dhiman, D. Deb, V. Muresan, V.E. Balas, Wake management in wind farms: an adaptive control approach, Energies 12 (7) (2019).

[2] J. Hattendorf, R. Low, W.-M. Gnehr, L. Wulff, M. Koekten, D. Koshnicharov, A. Blauaermel, J. Esquivel, Potential induced degradation in mono-crystalline silicon based modules: an acceleration model.

[3] P. Hacke, S. Spataru, K. Terwilliger, G. Perrin, S. Glick, S. Kurtz, J. Wohlgemuth, Accelerated testing and modeling of potential-induced degradation as a function of temperature and relative humidity, IEEE Journal of Photovoltaics 5 (6) (2015) 1549–1553.

[4] P. Hacke, K. Terwilliger, S. Glick, G. Tamizhmani, S. Tatapudi, C. Stark, S. Koch, T. Weber, J. Berghold, S. Hoffmann, M. Koehl, S. Dietrich, M. Ebert, G. Mathiak, Interlaboratory study to determine repeatability of the damp-heat test method for potential-induced degradation and polarization in crystalline silicon photovoltaic modules, IEEE Journal of Photovoltaics 5 (1) (2015) 94–101.

[5] P. Hacke, K. Terwilliger, R. Smith, S. Glick, J. Pankow, M. Kempe, S.K.I. Bennett, M. Kloos, System voltage potential-induced degradation mechanisms in PV modules and methods for test, in: 2011 37th IEEE Photovoltaic Specialists Conference, 2011, pp. 000814–000820.

[6] M. Schutze, M. Junghanel, M.B. Koentopp, S. Cwikla, S. Friedrich, J.W. Muller, P. Wawer, Laboratory study of potential induced degradation of silicon photovoltaic modules, in: 2011 37th IEEE Photovoltaic Specialists Conference, 2011, pp. 000821–000826.

[7] S. Pingel, O. Frank, M. Winkler, S. Daryan, T. Geipel, H. Hoehne, J. Berghold, Potential induced degradation of solar cells and panels, in: 2010 35th IEEE Photovoltaic Specialists Conference, 2010, pp. 002817–002822.

[8] P. Hacke, Characterization of multicrystalline silicon modules with system bias voltage applied in damp heat, in: 25th EUPVSEC, 2010, pp. 3760–3765.

[9] M. Schutze, M. Junghanel, O. Friedrichs, R. Wichtendahl, M. Scherff, J. Mueller, P. Wawer, Investigations of potential induced degradation of silicon photovoltaic modules, in: 26th EU PVSEC, 2011, pp. 3097–3102.

[10] D. Deb, N.L. Brahmbhatt, Review of yield increase of solar panels through soiling prevention, and a proposed water-free automated cleaning solution, Renewable & Sustainable Energy Reviews 82 (2018) 3306–3313.

[11] S. Mishra, K. Bhargava, D. Deb, Numerical simulation of potential induced degradation (pid) in different thin-film solar cells using scaps-1d, Solar Energy 188 (2019) 353–360.

[12] B. Espinar, J.-L. Aznarte, R. Girard, A.M. Moussa, G. Kariniotakis, Photovoltaic forecasting: a state of the art, in: 5th European PV-Hybrid and Mini-Grid Conference,

OTTI – Ostbayerisches Technologie-Transfer-Institut, Tarragona, Spain, ISBN 978-3-941785-15-1, 2010, pp. 250–255.
[13] V. Lara-Fanego, J. Ruiz-Arias, D. Pozo-Vázquez, F. Santos-Alamillos, J. Tovar-Pescador, Evaluation of the WRF model solar irradiance forecasts in Andalusia (southern Spain), Solar Energy 86 (8) (2012) 2200–2217, progress in Solar Energy 3.
[14] A. Moreno-Munoz, J.J. de la Rosa, R. Posadillo, F. Bellido, Very short term forecasting of solar radiation, 2008, pp. 1–5.
[15] D. Anderson, M. Leach, Harvesting and redistributing renewable energy: on the role of gas and electricity grids to overcome intermittency through the generation and storage of hydrogen, Energy Policy 32 (14) (2004) 1603–1614.
[16] T.R. Ayodele, A.S.O. Ogunjuyigbe, C.G. Monyei, On the global solar radiation prediction methods, Journal of Renewable and Sustainable Energy 8 (2) (2016) 023702.
[17] C. Osterwald, T. McMahon, J. del Cueto, Electrochemical corrosion of sno2:f transparent conducting layers in thin-film photovoltaic modules, Solar Energy Materials and Solar Cells 79 (2003) 21–33.
[18] P. Hacke, P. Burton, A. Hendrickson, S. Spataru, S. Glick, K. Terwilliger, Effects of photovoltaic module soiling on glass surface resistance and potential-induced degradation, 2015, pp. 1–4.
[19] S. Pingel, S. Janke, J. Seydewitz, R. Alam, S. Koch, J. Kupke, J. Berghold, The local potential distribution as driver of pid & "live pid monitoring" method, 2014.
[20] N. Shiradkar, E. Schneller, N.G. Dhere, Finite element analysis based model to study the electric field distribution and leakage current in PV modules under high voltage bias 8825 (2013) 88250G.
[21] S. Hoffmann, M. Koehl, Effect of humidity and temperature on the potential-induced degradation, Progress in Photovoltaics: Research and Applications 22 (2) (2014) 173–179.
[22] R.H. Inman, H.T. Pedro, C.F. Coimbra, Solar forecasting methods for renewable energy integration, Progress in Energy and Combustion Science 39 (6) (2013) 535–576.
[23] J.R. Andrade, R.J. Bessa, Improving renewable energy forecasting with a grid of numerical weather predictions, IEEE Transactions on Sustainable Energy 8 (4) (2017) 1571–1580.
[24] J. Berghold, S. Koch, A. Böttcher, A. Ukar, M. Leers, P. Grunow, Potential-induced degradation (pid) and its correlation with experience in the field, Photovoltaics International (2013) 87–93.
[25] K. Drabczyk, G. Kulesza-Matlak, A. Drygała, M. Szindler, M. Lipiński, Electroluminescence imaging for determining the influence of metallization parameters for solar cell metal contacts, Solar Energy 126 (2016) 14–21, https://doi.org/10.1016/j.solener.2015.12.029.
[26] K. Bedrich, W. Luo, M. Pravettoni, D. Chen, Y. Chen, W. Zigang, P. Verlinden, P. Hacke, Z. Feng, J. Chai, Y. Wang, A. Aberle, Y.S. Khoo, Quantitative electroluminescence imaging analysis for performance estimation of pid-influenced PV modules, IEEE Journal of Photovoltaics (2018) 1–8.
[27] C. Buerhop, J. Adams, S. Wrana, M. Dalsass, C. Zetzmann, C. Camus, C. Brabec, Ir-images of PV-modules with potential induced degradation (pid) correlated to monitored string power output, 2016.
[28] M. Florides, G. Makrides, G. Georghiou, Early potential induced degradation (pid) detection in the field: voltage measurement methods, 2017.
[29] C. Bedin, A. Oliveira, L. Nascimento, G. Xavier De Andrade Pinto, L. Augusto Zaniconki Sergio, R. Rüther, Pid detection in crystalline silicon modules using low-cost electroluminescence images in the field, 2018.
[30] F. Martinez Moreno, E. Lorenzo Pigueiras, J. Muñoz Cano, R. Parra, T. Espino, On-site tests for the detection of potential induced degradation in modules.

[31] R. Ghasempour, A. Puia, M. Ebrahimi, M. Ahmadi, H. Hadiyanto, Multi-criteria decision making (mcdm) approach for selecting solar plants site and technology: a review, International Journal of Renewable Energy Development 8 (2019) 15.
[32] A. Afshari, M. Mojahed, R. Yusuff, Simple additive weighting approach to personnel selection problem, International Journal of Innovation, Management and Technology 1 (2010) 511–515.
[33] B. Roy, The outranking approach and the foundations of electre methods, Theory and Decision 31 (1) (1991) 49–73.

CHAPTER 9

Next generation photovoltaics

The efficiency of the solar cells that are available in market today is still not satisfactory, and so there is a continuous endeavor to develop materials that may serve the purpose better. The challenge is the identification of newer processes of harvesting solar energy in highly efficient and cost-effective ways and also to create a valuable manufacturing proposition. The commercialization prospects of photovoltaics depend on the technological growth of next-generation photovoltaic cells and modules. They have shown a tremendous potential to replace the commercially available toxic and expensive photovoltaic technology globally. There are several alternatives available that could potentially be the PV technology of the future. However, several issues are hindering their progress from being the next generation PV. Therefore, it becomes essential for readers to become acquainted with the prospective next generation of PVs, the commercialization challenges involved, and the status of current research. This chapter presents a glimpse of the future of upcoming PV technologies, their research status, and market forecast.

To reach the full potential of PV deployment globally, there are many challenge that must be overcome: (1) The total cost of PV systems must be reduced while simultaneously increasing energy-conversion efficiency, reliability and lifetime; (2) the capacity to cost-effectively manufacture solar cells and balance the systems' components must be significantly increase; (3) methods to integrate PV into the utility grid must continue to improve to ensure affordable and reliable electricity; (4) transportation and the heating and cooling of buildings must be increasingly electrified; and (5) technologies must be improved for energy storage, catalyzed chemicals and fuels production, and water purification. In the rest of this book, we focus on the more recent solar cells that would facilitate progress in this direction. The solar panel-cleaning robots and self-cleaning mechanisms already described in this book, will remain important, despite the emergence of new materials for solar cells.

Degradation, Mitigation, and Forecasting Approaches in Thin Film Photovoltaics
https://doi.org/10.1016/B978-0-12-823483-9.00020-6

9.1 Recently developed photovoltaics

Humanity has witnessed spectacular progress of photovoltaic (PV) technology as a sustainable source of energy that is partly attributed to the alarmingly growing concerns over global warming and climate change. Moreover, the technological advancement of PV manufacturing has served the purpose [1]. The classification of PV generations (GENs) consists of four main categories, of which the first two generations currently dominate the PV market share. GEN-1 belongs to mono-/poly-crystalline silicon and gallium arsenide cells, which offer moderate efficiency levels at medium manufacturing costs. GEN-2 comprises the thin-film PV cells with lower manufacturing costs and lower conversion efficiency. GEN-3 cells are based on some novel absorber materials, and they offer low cost and flexible manufacturing with reasonable efficiency. GEN-4, popularly known as "inorganics-in-organics," combines the benefits obtained from GEN-2 and GEN-3 materials; however, these generations of PV cells are far from commercial success due to some other issues such as stability.

Most of the recently developed PVs belong to GEN-3 and GEN-4. The significant advantages of GEN-3 cells are solution-processability, large-area fabrication, mechanical flexibility, and reasonably high conversion efficiency. GEN-4 cells offer most of the benefits provided by GEN-3 cells, along with some unique ones like stability and better carrier separation. The authors have classified the GEN-3 and GEN-4 cells as the next generation PVs because they are still undergoing extensive laboratory research and far from commercialization but have shown immense potential to achieve the same soon. Some of the most promising next-generation PVs are discussed in the following sections.

9.1.1 Kesterite photovoltaics

Cu_2Z or kesterite (CZTS) PV technology is the first among the prospective candidates for next-generation PVs based on earth-abundant kesterite materials [2]. In a short period, kesterite PVs have shown an enormous potential to surmount the CIGS PVs market share. Such progress is attributable to the kesterite materials' special features, such as wide availability in nature, low processing cost, nontoxicity, decent optoelectronic properties, and stability [3]. There has been significant research progress made in the direction of the successful commercialization of kesterite PVs. The maximum efficiency of 12.6%, achieved in 2013, has become stagnant due to the emergence of more promising perovskite PVs.

However, a significant amount of research into this technology is still ongoing. While most investigations on kesterites aim to develop solar PV cells, alternative applications are also being explored. The lucrative features of kesterite materials can answer the issues related to environmental degradation and climate change, which is why the last two decades have witnessed some great contributions related to the development of kesterite PV technology. Apart from CZTS associated with two promising members of the kesterite family are $Cu_2ZnSnSe_4$ (CZTSe) and $Cu_2ZnSn(S,Se)_4$ (CZTSSe). All these materials demonstrate optoelectronic properties quite similar to CIGS materials. However, despite some great efforts, kesterite PVs' performance is still far from reaching its full potential. The reason behind this shortcoming lies in kesterites' inferior properties, such as narrow phase stability, deep-level defect concentration, and nonideal cell architecture. However, commendable research is still going on to overcome the issues hindering kesterite PVs' technological progress. Moreover, there have been substantial contributions exploring kesterites' utility beyond PVs, such as in charge-transport layers, water splitting, thermoelectric sensors, among other applications.

The crystal structure of kesterite is quite similar to chalcopyrite, i.e., zinc blende structure. Several complex compounds can be easily derived from the binary zinc blende structure merely by substituting one of the components with another element without disturbing the charge neutrality. Experimental findings have revealed that the tetragonal kesterite crystal is the most stable among the kesterite family. The bandgap of kesterite materials can be tuned within the range 1–1.5 eV by altering the S/Se ratio [4].

9.1.2 Dye-sensitized photovoltaics

Dye-sensitized photovoltaics (DSPVs) is one of the most easy-to-implement photovoltaic technologies, offering quite a reasonable conversion efficiency with low manufacturing cost and minimal material wastage [5]. They constitute the most efficient third-generation solar technology available, absorbing more sunlight per surface area than standard silicon-based solar panels. They are an attractive replacement for current technologies in low-density applications. The light weight and mechanical robustness are key benefit. Most of the materials used in DSPVs are less susceptible to contamination and are processable under room-temperature conditions. Moreover, DSPV materials such as titanium oxide are less prone to contamination, inexpensive, readily available (from natural sources), and environmentally friendly [6]. Due to these lucrative features, DSPVs can be manufactured on a large scale using a roll-to-roll process. Interestingly, they perform reasonably well under low-light intensities, making them favorable candidates for indoor applications [7].

The indoor application of DSPV modules can be made attractive by including advanced molecular engineering to incorporate colorful and transparent thin films during their manufacturing. Despite the features mentioned, DSPVs have not attracted the deserved research interest in the light of the availability of better and more efficient alternatives. The research in DSPVs has grown considerably to achieve a conversion efficiency of up to 13%. The typical structure of a DSPV cell consists of four main parts: a mesoporous oxide layer coated conductive glass substrate used as photo-anode; a monolayer of dye-sensitizer covalently bonded over the surface of the oxide layer (light absorber); an electrolyte having redox couple in an organic solvent (electron collector), and a counter electrode of a platinum-coated conductive glass substrate.

9.1.3 Organic photovoltaics

So far, we have discussed solar cells produced from inorganic materials like silicon, but researchers have begun investigating the use of cheap, earth-abundant, and safe organic compounds that have demonstrated the photovoltaic effect. Organic solar cells can be defined by their absorber material, which consists of either carbon-based conductive organic polymers or organic molecules. The currently popular organic photovoltaics (OPVs) are mostly based on bulk-heterojunctions (BHJs) used as active layers. Since organic semiconductors' discovery, tremendous progress has been made towards achieving the inexpensive, flexible, biodegradable, and easily processable OPVs [8].

The growth of OPVs has fuelled the expansion of the renewable energy share of solar photovoltaics. However, some technical issues have hampered their commercialization prospects [9]. The below-par conversion efficiency and low stability of OPVs have been the most pursued research problems for over a decade. The root cause of low efficiency is the morphological degradations occurring in BHJ thin films, disturbing the process of carrier generation, transport, and collection in OPV cells. The limitations imposed by BHJ thin films' poor morphology results in enhanced recombination rate at the sites present inside the grain boundaries. Thus, fine control over morphological tuning is significantly required to carry forward the progress of OPV technology. A unique feature of OPVs is that they can be printed on large-area based flexible substrates.

Additionally, most organic semiconductors are biodegradable, thus offering ecofriendly manufacturing with no recycling requirement. The OPV technology is also unique from the perspective of manufacturing the building-integrated photovoltaics (BIPVs) where transparency, flexibility, and light weight are critically needed [10]. Moreover, OPVs are a highly welcome addition to space programs where easy integration of PV modules is prioritized. OPVs utilize both small molecules and conducting polymers as the active layer materials since they demonstrate an ample sunlight-absorbing capability. OPV cells are fabricated in several configurations, namely, single-layer, bilayer, and BHJ configurations. The fabrication of OPV cells is facilitated by the sequential deposition of conductors and semiconductors using the solution-processed techniques (spin-coating, dip-coating, etc.).

The conventional OPVs are based on the Bulk Heterojunction (BHJ) active layer, a blend of p-type conducting polymers (electron donor) and n-type small molecule fullerenes (electron acceptor) with a conversion efficiency reaching 12%. Fullerenes' long-term stability is challenged due to inferior mechanical durability, weak visible-light absorption, chemical oxidation, rapid molecular diffusion, questionable photo/thermal stability, and photo-induced dimerization. However, the stability issue of fullerene acceptors has been countered by incorporating nonfullerene acceptors and some polymeric acceptors [11].

9.1.4 Quantum-dot photovoltaics

Quantum-dot sensitized photovoltaics (QDPVs) have an excellent chance to be the next-generation PV. An important feature that distinguishes them from others is the flexibility of size and compositional tunability for a wider solar spectrum [12]. The conversion efficiency of QDPV cells is close to 12%, which is on par with DSPVs. An essential aspect of QDPV cell operation is the generation of multiple numbers of electron-hole pairs upon a single-photon incidence. However, like DSPVs, its efficiency is also far below that achieved with single-junction PV cells, and therefore great research opportunities lie in this area.

The invention of QDPVs was motivated by efforts to replace molecular dyes with nanostructured semiconductors as sunlight absorbers. Thus, the configuration of the QDPVs cell is quite similar to DSPVs [13]. The nanostructured absorber layer's significance in QDPV cells is the larger surface area facilitating an increased absorption of photons. Quantum dots typically have a tiny feature size (3D confinement) comparable to an excitonic radius. The band-gap tunability in quantum dots is possible in a unique way by manipulating their size or alloying. The QDPV cell operation is prone to degradations caused by prolonged exposure to high oxygen and humidity levels and light and temperature [14].

9.1.5 Perovskite photovoltaics

A rapidly emerging PV technology is that of perovskite solar cells. Perovskites have seen a tremendous increase in initial efficiency in recent years. Perovskite cells with 2.2% were reported in 2006. Cells with 6.5% were reported in 2011, and since then the reported efficiencies have rapidly increased with an actual certified record efficiency of 19.7%. Uncertified cell efficiencies higher than 22% have been reported as well. In 2019, this 'special organic-inorganic halide photovoltaic material' experienced its tenth anniversary, but since then, its popularity as an optoelectronic material has grown many times over.

The most promising application of perovskite materials is the PV application as it has demonstrated outstanding performance in parallel with a relative ease of fabrication. Researchers have started hailing it as the most promising future PV material [15]. Perovskites are a class of materials with a crystal structure similar to calcium titanium oxide ($CaTiO_3$) and strong optoelectronic properties. The initially reported efficiency of 3.8% has advanced to 25.2%, which is on par with other commercial PV cells. The thin-film perovskite absorbers offer a high absorption coefficient ($>10^5$ cm^{-1}), long carrier-diffusion length (>1 μm), high and balanced electron and hole-mobility values, low exciton-binding energy (<10 meV), long carrier lifetime (>300 ns), optimum and tunable bandgap, better crystal defect tolerance, and low surface-recombination velocity [16].

The perovskite materials are low-cost and easily made into thin-films under ambient conditions using simple spin-coating, screen printing, dip coating, etc. Moreover, its deposition is fully compatible with flexible substrates. Perovskites can be considered a potential PV material equipped with the advantages of both silicon and organic materials. However, some demerits of perovskite PVs, such as poor stability and short lifetime, are currently the major obstacles on the path to successful commercialization.

The conversion efficiency of perovskite PVs lag behind silicon PVs. However, recently, the combination of silicon and perovskite has shown better efficiency than silicon PVs. In the frequently used perovskite PV material $CH_3NH_3PbI_3$, however, the lead content poses a severe threat to the environment and suffers from poor stability. Thus, researchers around the globe are now working towards the development of lead-free perovskite PVs. In this regard, the PV potential of $CH_3NH_3SnI_3$, $CH(NH)_2SnI_3$, and $CsSnI_3$ perovskite materials are now being explored rigorously [17,18].

9.2 Commercialization challenges for next-generation PVs

The various next-generation PV technologies discussed in the previous section are promising but are currently facing severe technological challenges impeding their growth towards successful commercialization. The future growth prospects of next-generation PV are strongly dependent on the development of materials for use in manufacturing. In this section, we present a summary of the technological challenges faced by emerging PVs. The summary is based on the aspects related to both materials and devices. The discussion is expected to provide a quick overview of the current challenges and research opportunities in this area. The details of the issues pertaining to the next-generation PVs accessed through the references cited in Table 9.1.

9.3 Current research status and forecast for next-generation PVs

The next-generation PV technologies are currently undergoing some significant laboratory-based development, and we expect commercial viability soon. Most of the upcoming PV technologies, such as DSPVs, OPVs, and perovskite PVs, are expected to be low-cost, large-area-based, and ecofriendly. The next generation PVs require novel materials (semiconductors, conductors, encapsulants, and substrates), and they also suffer from low reproducibility of their performances. Therefore, research on emerging PVs aims to achieve highly efficient, low-cost manufacturing, and stable and environment-friendly technology. However, the investigation's target is still unclear as to what level of technology we can achieve. The development of next-generation PVs depends on the growth of the PV market and research methods.

The growth of science and technology is desirable for the subsequent development of PV technology. From this perspective, the synthesis and characterization of advanced PV materials are essential. An enormously large-scale growth of advanced PV modules and systems will require inno-

Table 9.1 Summary of commercialization challenges in next-generation PVs.

Photovoltaics	Commercialization challenges
Kesterite [19]	(i) Requirement of off-stoichiometric composition (ii) Decomposition at high temperature (iii) Compositional/processing-induced intrinsic defects (iv) Deficit of 'V_{OC}'
Dye-sensitized [20]	(i) Availability of dye materials (ii) High recombination rate (iii) Inferior lifetime (iv) Protection of metal interconnects between cells
Organic [21]	(i) Selection of donor and acceptor materials (ii) Low photo-/thermal-stability (iii) Poor active-layer morphology (iv) Low power-conversion efficiency (v) Degradation due to oxygen, moisture and UV radiation
Quantum-dot [22]	(i) Low stability (ii) Toxicity (iii) Robustness (iv) Quantum-dot loading enhancement of photoanodes (v) Requirement of advanced electrodes
Perovskite [23]	(i) Toxicity (ii) Thermal and UV degradation (iii) Low chemical stability (iv) Ion migration (v) Poor interfacial properties

vations related to minimization of losses, better control strategies, developed and compact power electronics, smarter temperature control, long-term stability, better storage capability, improved grid compliance, and accurate maximum power-point tracking. Hence, we need to systematically deal with the challenges of improving future PV systems' commercialization prospects through scientific and technological innovations. The growth forecast for next-generation PVs during 2020–2040 is dependent on the emergence of unique materials. The upcoming trend of photovoltaics is now trending in the direction of flexible photovoltaics. Thus the focus of the research community is shifting towards the development of unique stretchable materials.

The novel applications, like mobile phones, aircraft, internet-of-things (IoT) nodes, BIPVs, etc., require the development of flexible PV modules. We expect renewable energy-based electricity generation to grow to more than 920 GW by 2022. With various new energy startups, the implementation of PV power projects is progressing at a rapid speed. The consumer response to PV power is growing each year because it is now available at affordable rates. With increasing competition between companies, we expect the prices to dip further. Significantly, developing countries like India and China are front-runners in PV innovations.

Bibliography

[1] M.H. Shubbak, Advances in solar photovoltaics: technology review and patent trends, Renewable & Sustainable Energy Reviews 115 (2019) 109383, https://doi.org/10.1016/j.rser.2019.109383.

[2] A.S. Nazligul, M. Wang, K.L. Choy, Recent development in earth-abundant kesterite materials and their applications, Sustainability 12 (12) (2020) 5138, https://doi.org/10.3390/su12125138.

[3] S. Giraldo, Z. Jehl, M. Placidi, V. Izquierdo-Roca, A. Pérez-Rodríguez, E. Saucedo, Progress and perspectives of thin film kesterite photovoltaic technology: a critical review, Advanced Materials 31 (16) (2019) 1806692, https://doi.org/10.1002/adma.201806692.

[4] S. Mohammadnejad, A.B. Parashkouh, CZTSSe solar cell efficiency improvement using a new band-gap grading model in absorber layer, Applied Physics A 123 (12) (2017), https://doi.org/10.1007/s00339-017-1371-x.

[5] J. Gong, K. Sumathy, Q. Qiao, Z. Zhou, Review on dye-sensitized solar cells (DSSCs): advanced techniques and research trends, Renewable & Sustainable Energy Reviews 68 (2017) 234–246, https://doi.org/10.1016/j.rser.2016.09.097.

[6] N. Mariotti, M. Bonomo, L. Fagiolari, N. Barbero, C. Gerbaldi, F. Bella, C. Barolo, Recent advances in eco-friendly and cost-effective materials towards sustainable dye-sensitized solar cells, Green Chemistry 22 (21) (2020) 7168–7218, https://doi.org/10.1039/d0gc01148g.

[7] S. Biswas, H. Kim, Solar cells for indoor applications: progress and development, Polymers 12 (6) (2020) 1338, https://doi.org/10.3390/polym12061338.

[8] H. Gaspar, F. Figueira, L. Pereira, A. Mendes, J. Viana, G. Bernardo, Recent developments in the optimization of the bulk heterojunction morphology of polymer: fullerene solar cells, Materials 11 (12) (2018) 2560, https://doi.org/10.3390/ma11122560.

[9] G. Chidichimo, L. Filippelli, Organic solar cells: problems and perspectives, International Journal of Photoenergy 2010 (2010) 1–11, https://doi.org/10.1155/2010/123534.

[10] B. Lee, L. Lahann, Y. Li, S.R. Forrest, Cost estimates of production scale semitransparent organic photovoltaic modules for building integrated photovoltaics, Sustainable Energy & Fuels 4 (11) (2020) 5765–5772, https://doi.org/10.1039/d0se00910e.
[11] C. Yan, S. Barlow, Z. Wang, H. Yan, A.K.-Y. Jen, S.R. Marder, X. Zhan, Non-fullerene acceptors for organic solar cells, Nature Reviews Materials 3 (3) (2018), https://doi.org/10.1038/natrevmats.2018.3.
[12] A. Kongkanand, K. Tvrdy, K. Takechi, M. Kuno, P.V. Kamat, Quantum dot solar cells. Tuning photoresponse through size and shape control of CdSe-TiO2 architecture, Journal of the American Chemical Society 130 (12) (2008) 4007–4015, https://doi.org/10.1021/ja0782706.
[13] A. Sahu, A. Garg, A. Dixit, A review on quantum dot sensitized solar cells: past, present and future towards carrier multiplication with a possibility for higher efficiency, Solar Energy 203 (2020) 210–239, https://doi.org/10.1016/j.solener.2020.04.044.
[14] Y. Chen, Y. Zhao, Incorporating quantum dots for high efficiency and stable perovskite photovoltaics, Journal of Materials Chemistry A 8 (47) (2020) 25017–25027, https://doi.org/10.1039/d0ta09096d.
[15] P. Roy, N.K. Sinha, S. Tiwari, A. Khare, A review on perovskite solar cells: evolution of architecture, fabrication techniques, commercialization issues and status, Solar Energy 198 (2020) 665–688, https://doi.org/10.1016/j.solener.2020.01.080.
[16] S.G. Mhaisalkar, N. Mathews, H.J. Bolink, D. Bahulayan, Advances in perovskite optoelectronics: bridging the gap between laboratory and fabrication, Advanced Energy Materials 10 (13) (2020) 2000393, https://doi.org/10.1002/aenm.202000393.
[17] A. Abate, Perovskite solar cells go lead free, Joule 1 (4) (2017) 659–664, https://doi.org/10.1016/j.joule.2017.09.007.
[18] V. Pecunia, L.G. Occhipinti, A. Chakraborty, Y. Pan, Y. Peng, Lead-free halide perovskite photovoltaics: challenges, open questions, and opportunities, APL Materials 8 (10) (2020) 100901, https://doi.org/10.1063/5.0022271.
[19] M. He, C.Y.J. Li, M.P. Suryawanshi, J. Kim, M.A. Green, X. Hao, Kesterite solar cells: insights into current strategies and challenges, Advanced Science 8 (9) (2021) 2004313, https://doi.org/10.1002/advs.202004313.
[20] J.B. Baxter, Commercialization of dye sensitized solar cells: present status and future research needs to improve efficiency, stability, and manufacturing, Journal of Vacuum Science & Technology A: Vacuum, Surfaces, and Films 30 (2) (2012) 020801, https://doi.org/10.1116/1.3676433.
[21] A. Bonasera, G. Giuliano, G. Arrabito, B. Pignataro, Tackling performance challenges in organic photovoltaics: an overview about compatibilizers, Molecules 25 (9) (2020) 2200, https://doi.org/10.3390/molecules25092200.
[22] I. Mora-Seró, Current challenges in the development of quantum dot sensitized solar cells, Advanced Energy Materials 10 (33) (2020) 2001774, https://doi.org/10.1002/aenm.202001774.
[23] L. Qiu, L.K. Ono, Y. Qi, Advances and challenges to the commercialization of organic–inorganic halide perovskite solar cell technology, Materials Today Energy 7 (2018) 169–189.

Index

K

L

M

N

O

P

W